ECOSYSTEMS & SPECIES
The Web of Nature

ROE DEER PEERING ABOVE THICK VEGETATION

Dr. Timothy Polnaszek is an Assistant Professor at Belmont Abbey College. Previously, he worked at MacMurray College and was a postdoctoral research fellow at the University of Arizona within the Center for Insect Science. Dr. Polnaszek obtained his BS in Zoology from North Dakota State University and his PhD from the University of Minnesota, where he conducted research on animal behavior. Dr. Polnaszek's research projects have included the decision-making of animals such as woodchucks, blue jays, bumble bees, crickets and others. He has presented this research at national and international conferences, as well as published journal articles on wildlife and animal behavior. He now studies animals and plants on a small forest plot in Gastonia, NC that he calls home, as an ongoing research collaboration with his wife and six children.

ECOSYSTEMS & SPECIES
The Web of Nature

Timothy Polnaszek, PhD

TAN Books
Gastonia, North Carolina

Ecosystems & Species: The Web of Nature © 2023 Timothy Polnaszek, PhD

All rights reserved. With the exception of short excerpts used in critical review, no part of this work may be reproduced, transmitted, or stored in any form whatsoever, without the prior written permission of the publisher. Creation, exploitation and distribution of any unauthorized editions of this work, in any format in existence now or in the future—including but not limited to text, audio, and video—is prohibited without the prior written permission of the publisher.

Unless otherwise noted, Scripture quotations are from the Revised Standard Version of the Bible—Second Catholic Edition (Ignatius Edition), copyright © 2006 National Council of the Churches of Christ in the United States of America. Used by permission. All rights reserved.

Excerpts from the English translation of the Catechism of the Catholic Church for use in the United States of America © 1994, United States Catholic Conference, Inc.—Libreria Editrice Vaticana. Used with permission.

All excerpts from papal homilies, messages, and encyclicals Copyright © Libreria Editrice Vaticana. All rights reserved.

Cover & interior design and typesetting by www.davidferrisdesign.com

ISBN: 978-1-5051-2866-6

Published in the United States by
TAN Books
PO Box 269
Gastonia, NC 28053

www.TANBooks.com

Printed in the United States of America

> "Where were you when I laid the foundation of the earth?"
>
> —Job 38:4

GREEN SEA TURTLE

CONTENTS

Preface VIII

Introduction XI

Chapter 1: Ecology and Ecosystems: Studying the Organization of Nature 1

Chapter 2: Adaptations: How Organisms Deal with Extreme Temperatures 9

Chapter 3: Adaptations: The Challenges of Water Balance 17

Chapter 4: Adapting to the Environment: How Populations Change over Time 27

Chapter 5: Natural Selection: How Populations Evolve and Change 37

Chapter 6: Population Biology: How Population Size Changes over Time 47

Chapter 7: Competition: Conflicts, Contests, and Battles in Nature 57

Chapter 8: Exploiting Other Species: Predation, Herbivory, and Parasitism 67

Chapter 9: Teamwork in Nature: Social Groups, Cooperation, and Mutualisms 77

Chapter 10: Communities: Food Webs, Connections, and Cascades 85

Chapter 11: Biological Communities and Biodiversity 93

Chapter 12: Ecosystems: Finding Balance within Communities and Environments 101

Conclusion 107

Amazing Facts about Ecosystems and Species 108

Key Terms 118

PREFACE

When I think about the scientific study of the natural world, two phrases from the writings of Pope St. John Paul II come to mind:

(1) a rigorous pursuit of truth and

(2) a love of learning.

The first—a rigorous pursuit of truth—describes science and its processes. Scientists make careful observations, design experiments, and collect data to learn more about how the world works. Too often, though, science may seem like something people do in a big research facility while wearing lab coats.

But we are all scientists!

Anyone can study the living world in a scientific way. From an early age, each of us has a curiosity to understand the world. Think of babies repeatedly dropping something onto the floor; they are discovering how gravity works! It is this basic curiosity that drives science.

The second piece—a love of learning—also describes what science should inspire. Sometimes science is depicted as a dry, boring set of facts, but nothing could be further from the truth. The world is a fascinating place. I have been interested in the natural world my whole life. This love of nature led me to obtain undergraduate and postgraduate degrees that have allowed me to teach biology classes every day for a living, and yet I am still constantly amazed by the wonders of our world.

There is always something new to learn in biology and all the natural sciences. Within biology, there is so much inspiring beauty in the endless forms continually evolving and unfolding. Life on Earth consistently exceeds the capacity of my imagination.

For example, did you know:

WOOD FROGS

- Like animals, plants can "sweat"? They release water through small pores in the leaf tissue called stomata, and the evaporation of this water cools the plant during periods of high temperatures.

- Wood frogs live throughout Canada and Alaska because they can survive even frozen solid in winter (like a frog Popsicle) and simply thaw out in warmer months?

- In addition to camouflage or disruptive coloration, organisms also use mimesis to look like something uninteresting or non-threatening, like a leaf or twig, or mimicry to look like something dangerous—like caterpillars that have eyespots to resemble a snake?

How could we not help but love to learn about this fascinating world we inhabit?

> *"[Science and faith] each can draw the other into a wider world, a world in which both can flourish."*
>
> —Pope St. John Paul II in *Physics, Philosophy and Theology*

Finally, it is too often assumed in society today that faith and science act in opposition to one another, that somehow if we learn enough about the world, what we learn would disprove the existence of God. But it is important for each of us to be confident in our Faith and in the fact that truth cannot be in opposition with itself.

We read in the *Catechism of the Catholic Church*: "Methodical research in all branches of knowledge, provided it is carried out in a truly scientific manner and does not override moral laws, can never conflict with the faith, because the things of the world and the things of faith derive from the same God. The humble and persevering investigator of the secrets of nature is being led, as it were, by the hand of God in spite of himself, for it is God, the conserver of all things, who made them what they are" (*CCC* 159).

Holy Mother Church teaches us that we can pursue scientific knowledge unafraid. It is my hope that *The Foundations of Science* series will not simply give your family some facts about the world but instead instill a curiosity and love of learning in you that you can apply across all the disciplines of your life, both scientific and otherwise.

Timothy Polnaszek, PhD

EUROPEAN BEE-EATER

INTRODUCTION

In this psalm, we hear about all of nature praising God. Saint Augustine discusses this passage by telling us that the beauty of creation is like a voice that sings out to confess God's greatness, but that nature *finds its voice through us*. The very beauty of all these things is like a voice that they raise to confess God. The sky cries to God, "You made me, I did not make myself." The earth cries out, "You founded me, I did not establish myself." How do these things cry out in worship? Whenever men and women observe them and discover the truth of them, all creatures cry out through people's appreciation of them; they shout with our voices.

So, when we gaze with wonder upon nature's beauty and through that wonder seek to understand it and discover the hidden truths buried out in the wild (that is, by being scientists!), we give glory to God. All creatures cry out with joy to their Creator when we marvel at them and appreciate them. They join their voices to ours in praise!

Learning about nature and how natural systems operate also provides us with the knowledge we need to better harness and safeguard the benefits the natural world provides us with (food, water, natural resources, beautiful places to appreciate, and so much more!). In other words, we can be better stewards of creation the better we understand it.

This book is filled with information about ecosystems—animals, plants, their relationships with one another, communities of organisms, the environment, the balance of nature, and more! As we will see, ecology is a field of biology that studies all of these things. It studies the physical environment and how animals and plants live within it. I hope this book generates a renewed interest in the natural world in you and your family, and that as you learn more about ecosystems you are filled with wonder and awe at God's creation, for in doing so, our wonder and awe brings greater glory to God!

"Praise the Lord from the earth,
you sea monsters and all deeps,
fire and hail, snow and frost,
stormy wind fulfilling his command!
Mountains and all hills,
fruit trees and all cedars!
Beasts and all cattle,
creeping things and flying birds!
Kings of the earth and all peoples,
princes and all rulers of the earth!
Young men and maidens together,
old men and children!
Let them praise the name of the Lord,
for his name alone is exalted;
his glory is above earth and heaven."

–Psalm 148:7–13

Yosemite Valley in California is one of many awe-inspiring landscapes found all over earth. Beneath the surface of this beauty lies an intricate balance that supports a complex web of life.

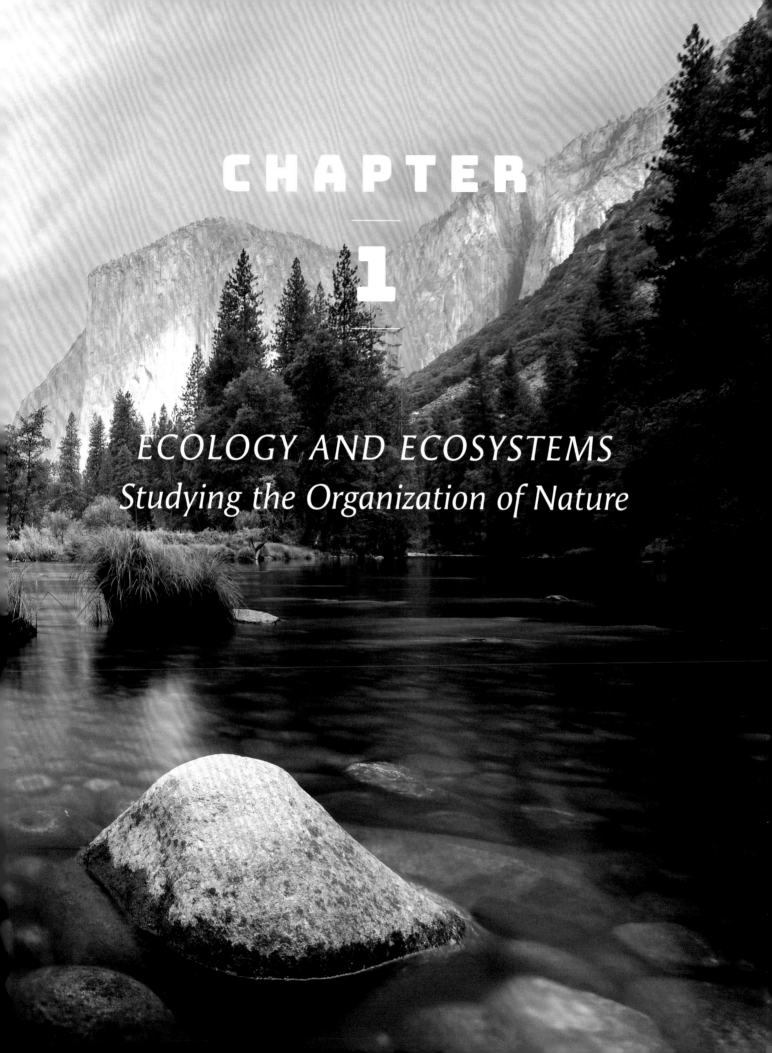

CHAPTER 1

ECOLOGY AND ECOSYSTEMS
Studying the Organization of Nature

Hawks are just one of many birds of prey that influence the population of rodents in a given ecosystem as they hunt and capture them for a nice meal.

WHAT IS ECOLOGY?

Ecology may be a less familiar term than, say, biology. As you may have heard before, *bio* means "life," so biology is the study of life. What then is ecology? Well, *eco* comes from the Greek word *oikos*, which means "household." In literal terms, then, **ecology** is the study of the home or household—which makes sense if we consider our home to mean our planet—Earth.

Another definition for ecology is the study of living things and their relationships with their surrounding environment. But when studying ecology, we don't just explore organisms and how they relate to the physical environment (things like temperature or rainfall), but also their *relationships to one another*. You may have heard the terms "food chain" and "food web" to describe which organisms eat other organisms. For example, hawks eat mice, which in turn eat seeds and grasses. This relationship between an individual predator (a hawk) and its prey (a mouse) is one of those relationships that would be of interest to **ecologists** (scientists who study ecology).

But ecologists would also consider how the number of mice living in a field might change if there were an increase (or decrease) in hawks nearby. Or they might consider how runoff from fertilizer could increase the food available to mice, which could then increase the number of mice. These are only two examples of the questions ecologists might ask, and that's only when thinking about hawks and mice and seeds. Obviously, there are many more questions when considering all of nature and the links between the many organisms.

One last way to describe ecology is to say that ecologists are interested in the "interconnectedness" of life on Earth. When we look at the natural world, we see a system of interdependent parts. In the previous example, the number of mice depends on the number of hawks, but also on the number of seeds. In turn, the number of seeds produced by plants would depend on the temperature and rainfall in the local area. This interconnectedness can make it difficult to study ecology because whatever part of nature you're interested in may depend on many hundreds of other factors. Ecologists tend to define different levels of organization to help with this problem. Let's take a look at some of these.

NATURE'S LEVELS OF ORGANIZATION

Individuals: Ecologists might study individual organisms—their behavior, physiology, lifespan, food, natural predators, diseases, and more. Topics would include how individuals respond to other living organisms or to their environment (temperature, moisture, light, chemicals, etc.). For example, we could ask how well a blue jay remembers where it buried acorns. Or, how does increasing average temperature affect the lifespan of a pika?

Populations: You've probably heard this word before. Population can refer to how many people live in a given city or town. For our purposes in ecology,

population is a group of individuals of the same species within a given area. Some ecologists study populations and how they change over time—this could mean **population dynamics** (is the population increasing or decreasing in number) or the adaptation of a population to its environment. For example, we could ask how annual rainfall affects the number of mosquitoes in an area. Or, how does this population of crickets change when a new type of predator moves into the area?

Species: Each distinct type of organism is called a **species.** There may be multiple populations of the same species spread out across a continent. A good example is white-tailed deer. They live across North America, from Canada to Mexico. We probably wouldn't say that a deer in Canada is in the same *population* as one in Mexico. But both deer living in Canada and those that live in Mexico are part of the same species. Rather than asking questions about a particular population, scientists might be interested in the whole species.

Each species has its own two-part scientific name: the first part is the genus and the second is the species. For white-tailed deer, this is *Odocoileus virginianus*. The mule deer is in the same genus—*Odocoileus*—but is a different species—*Odocoileus hemionus*. Some species have interesting names, including one jumping spider—*Indomarengo chavarapater*—named after St. Kuriakose Elias Chavara from India! When writing out these names, we use italics and capitalize the genus name, but not the species name. Throughout this book, we'll add in the scientific names of some of the animals and plants so that we become more familiar with this naming system.

This white-tailed deer buck (left) and mule deer buck (right) are of the same genus but are different species.

Community: Now that we have a definition of both population and species, we can look at communities. A **community** is the collection of populations of various species that interact with one another in a particular area. So the community in my backyard would include various bird species, squirrels, chipmunks, and mice that all interact to compete for seeds at or near my family's birdfeeder. We could study which species eats the most seed, and how that might change depending on what type of seeds we put out, how many seeds, or where we put them in the yard (i.e., on the ground versus in a tree).

LEVELS OF ORGANIZATION IN ECOLOGY

INDIVIDUAL

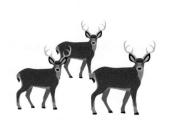

POPULATION/SPECIES

COMMUNITY

ECOSYSTEM

BIOME

BIOSPHERE

The community in my backyard would also include trees and other plants growing there, and hawks and owls that might prey on the seedeaters.

Ecosystem: In our definition of communities, there was no mention of the abiotic factors that affect living things. **Abiotic** means non-living (the prefix "a" means "non" and bio = life, so abiotic = non-living). An **ecosystem** involves the biological communities living in an area and how these communities (with all their populations of organisms) are affected by abiotic factors and the physical environment. So as we scale up in organization and levels of study, the ecosystem includes abiotic factors such as average temperature or rainfall or the amount of nutrients in the soil.

Biome: **Biomes** are particular types of ecosystems found in particular geographic regions. These biomes are defined, in large part, by average temperature and rainfall in the area. Examples of biomes include tropical rainforest, savanna, desert, and tundra. Each biome type has defining characteristics which all of the biomes of that type share. For example, the tropical rainforest ecosystems across the globe are not identical, but they all share similarities with one another (lots of rainfall and high temperatures year-round). Similarly, different species live in each rainforest, but each species would have specialized traits or adaptations that allow them to thrive in this biome.

Biosphere: The entirety of all the ecosystems on Earth form our planetary ecosystem, or **biosphere**. A forest in northern Minnesota or in Canada is quite far away from a rainforest in Indonesia or a desert in South Africa, such that these places seem unconnected. But there are ecological processes which we can study even at a global scale. Ecological relationships between living organisms and the abiotic environment affect the movement of water, nutrients

Clockwise from the top left we discover four extremely different examples of biomes that make up our biosphere: a river cutting through a gorge in the Wadi Mujib Biosphere Reserve, a tropical jungle in Costa Rica, the busy metropolis of Bangkok, Thailand, and a winter landscape in the Austrian Alps.

(like nitrogen or phosphorus), energy, and other factors. These interactions that occur at a local level can add up to effects at a global scale. For example, plants all over the planet, in each type of biome, help us by producing oxygen and absorbing carbon dioxide from the atmosphere.

THOSE WHO STUDY ECOLOGY

As we mentioned before, ecologists are those scientists who study ecology. These men and women may specialize on a particular level we just discussed or they may study questions that span multiple levels. Ecology is also a very interdisciplinary science, which means that many fields of science are important in helping us understand the relationships between organisms and the environment and ecological processes. Examples include physical sciences such as chemistry, meteorology, and atmospheric sciences, and biological sciences like physiology, genetics, and animal behavior. The scale of study for ecological research could also range from the microscopic (how microorganisms interact in a small scoop of soil) to global (how plant life in our forests affects the atmosphere on Earth).

In this book, we will explore these levels of organization in a little more detail. Understanding each level of nature's "system" can help us better understand the whole. We'll learn a lot along the way, but more importantly, I think we'll have a lot of fun on this adventure!

FOUNDATIONS REVIEW

✓ Ecology is the study of living things and their relationship with their surrounding environment. When studying ecology, we don't just mean organisms and how they relate to the physical environment (things like temperature or rainfall), but also their relationships to one another.

✓ The interconnectedness we see in nature can make it difficult to study ecology because whatever part of nature you're interested in may depend on many hundreds of other factors. Ecologists tend to define different levels of organization to help with this problem. These levels include: Individuals, Populations, Species, Communities, Ecosystems, Biomes, and Biospheres.

✓ Ecologists are scientists who study ecology. These men and women may specialize on a particular level or they may study questions that span multiple levels.

Ecology and Stewardship

In this book on ecology and ecosystems, we will see how God's creation is a dynamic, interconnected system, like how food webs rely on the connections between organisms, or how predators keep prey populations in balance. There are also times when this balance is thrown off, for example when invasive species invade a new community. As stewards of creation, and as scientists, we all can study ecosystems to better understand how to protect their beauty and biodiversity.

Several recent popes have highlighted this in their writings and teachings. Pope Benedict XVI once said, "The created world, structured in an intelligent way by God, is entrusted to our responsibility and though we are able to analyze it and transform it we cannot consider ourselves creation's absolute master. We are called, rather, to exercise responsible stewardship of creation, in order to protect it, to enjoy its fruits, and to cultivate it, finding the resources necessary for every one to live with dignity" (General Audience August 26, 2009).

In the encyclical *Centesimus Annus*, Pope Saint John Paul II discusses how natural systems have order, organization, and proper functioning (or that nature works in a balance). Finally, the USCCB tells us, "We show our respect for the Creator by our stewardship of creation. Care for the earth is not just an Earth Day slogan, it is a requirement of our faith. We are called to protect people and the planet, living our faith in relationship with all of God's creation." By working with the environment and respecting God's creation, we can grow closer to Him and better care for our brothers and sisters.

Pope Saint John Paul II designated Saint Francis of Assisi as the patron saint for ecology and ecologists—so we can ask for his intercession as you read through this book and learn more about these topics!

A lion enjoys shade in the sunbaked savannah of Africa. Behavioral adaptations, like hiding in the shade, are one way organisms deal with extreme temperatures in the wild.

CHAPTER 2

ADAPTATIONS
How Organisms Deal with
Extreme Temperatures

STARTING WITH INDIVIDUALS

We will start our exploration of the "levels" of ecology by first looking at individuals, before "scaling up" to populations, communities, and ecosystems. In this chapter, we will look closely at how individual organisms deal with environmental challenges. To survive in a given biome such as a rainforest, desert, or tundra, an organism must be able to handle the environment found there. There are many different abiotic factors that affect an organism's ability to survive in an area, and many unique and fascinating adaptations which help meet those abiotic challenges—too many to explore them all here. But we will examine a few of the major categories of abiotic challenges, starting with temperature, and then showcase a few specific examples of how animals, plants, fungi, or microorganisms deal with those factors.

SURVIVING THE HEAT AND THE COLD

One key abiotic factor that we're all familiar with is temperature. We experience changes in temperature across the seasons, or at least we do anytime we leave our climate-controlled homes and buildings. But other organisms do not have air conditioning or central heating, so how do they deal with fluctuations in temperature?

Some biomes can be extremely hot or extremely cold. The Sahara Desert has daily high temperatures over 100 degrees Fahrenheit (F) for several months during the summer. In contrast, the tundra in Canada is consistently below 0 degrees F during the winter—even reaching below -60 degrees F from time

The African desert and the Canadian tundra show just how drastic earth's temperature differences can be, with a disparity of almost 200 degrees! Organisms who live in these places face intense challenges to their survival.

to time! We wouldn't expect animals, plants, or fungi to be able to survive in both these places, that is, to survive in both the tundra *and* the desert—the swings in temperature are just too dramatic. Those that could live in one could never survive in the other. In fact, it is surprising that anything can live in either place with such extreme conditions. But organisms have special traits or adaptations to help them deal with extreme challenges in their habitats—like these extreme hot or cold conditions. The attributes they have fit with their particular habitats, but are often unsuited to different types of challenges. The process of **adaptation** describes the ways in which organisms become more suited to their environment over time.

ANIMAL ADAPTATIONS TO TEMPERATURE

Behavior, meaning the way animals act, is one way that animals cope with extreme temperatures. Some animals migrate and, in doing so, simply avoid extreme cold or hot months. For example, many North American bird species leave their homes and spend winters down in Mexico or other southerly places where it is warmer. Many other animals also change their behavior patterns depending on the temperature, even within a single day. In hot deserts, for example, animals may be more active at dusk and dawn or even at night, when it is not as hot, as they hunt and search for food.

Jaguars can hunt during the day, but are mostly active at night, which lets them escape the heat of the day.

These sorts of behaviors shouldn't be all that unfamiliar to us since we humans do similar things. We, too, change our behavior based on the temperature. During the hot summer months, we go swimming to cool off and wear lighter clothes, while in the winter we throw on coats when we go out in the snow. Many retired people, perhaps even your own grandparents, will live in different places throughout the year, traveling down to Florida or another southern state to avoid cold winters up north.

Responding to challenging temperatures isn't all about behavior, though. After all, plants and fungi can't move around to hide in the shade on hot summer afternoons. The physiology of an organism can also help it deal with extreme hot or cold conditions. **Physiology** is defined as all of the parts of an organism and the normal functioning of those parts. By "parts" of an organism we mean muscles, blood vessels, and nervous systems in animals, and leaves, stems, and roots in plants. The way these systems function can help an organism deal with temperatures. Sweating is one way that many mammals cool off, including us. The body releases sweat and, as it evaporates off the skin, that transition of water from liquid to gas decreases the temperature of the skin.

Remember:

Adaptations can be behavioral, like migrating or hibernating, or physiological, which means parts of the organism's body can be suited to environmental challenges, like carnivores having sharp teeth.

Staying Extra Cool

While sweating is one way mammals cool off, what if you live somewhere where sweating simply isn't enough? Some kangaroo species living in the hottest parts of Australia lick their forearms to add more cooling—the saliva evaporates and cools the area like sweat evaporating. Moreover, there are many blood vessels under the forearm, so the blood there cools off before returning to the heart and circulating elsewhere. In this way, licking their forearms can help them cool off their whole body!

Just as animals in warm places need to find ways to cool down, animals in cold ecosystems, like the tundra, need ways to stay warm. One primary way that animals stay warm is by using insulation. Just like we use blankets to stay warm at night to insulate us from the cold, animals have various special features which help them stay warm. You probably know that fur for mammals and feathers for birds help them keep warm, but did you know that some animals have layers of **blubber** (a type of fat) that also provide insulation? Elephant seals (*Mirounga leonina*), which sometimes swim in cold-water environments, have bodies that are up to 40 percent fat!

Animals have all sorts of ways to stay warm in frigid environments, from having fur and layers of blubber, to cuddling together to share body heat!

Some animals do not keep themselves warm, but use outside sources of heat to warm themselves—these are called **ectotherms** (the prefix *ecto* comes from the Greek *ektos*, meaning "outside," and *therm* means "heat," so "heat from outside"). Birds and mammals that use their own body heat to stay warm are **endotherms** (*endo* from the Greek for "inner" or "inside"). But ectotherms that can't warm themselves up from the inside might bask in the sun to warm up, which is why you see lizards and snakes lounging in the sun on warm rocks or bricks. Ectotherms use less energy because they don't use their own energy to raise their body temperature, but they may be limited in where they can live because they must depend on outside factors to stay warm.

Another way to survive through extreme cold is to wait in a resting state until the weather returns to a more comfortable temperature. Animals can do this through **hibernation**—storing up fat and remaining in a dormant, almost sleep-like state through the winter. Hibernation isn't just sleep, though; the

The artic fox's white coat not only helps keep it warm but helps it blend in with the snowy terrain, making the prey it stalks less likely to see it coming.

animal enters a deep state of physical resting. These animals have an extremely reduced metabolic rate (reduced breathing, heart rate, etc.)—they may only need about 5 percent of the energy they would typically need when awake!

Based on an animal's physical traits, we may be able to predict where they live (and likewise for plants, fungi, or other organisms). Because certain mammals have fur and other adaptations to stay warm, they can live in colder environments than many ectotherms. Arctic foxes, polar bears, caribou, and walruses all live near the Arctic and the North Pole. There are almost no lizards or snakes that live as far north as these mammals. However, in biology there are almost always exceptions. Wood frogs live throughout Canada and Alaska because they can survive even frozen solid in winter (like a frog Popsicle) and simply thaw out in warmer months!

PLANT ADAPTATIONS TO TEMPERATURE

Now we will see how some plants deal with extreme temperatures. Unsurprisingly, trees cannot fly south for the winter to find warmer weather. So how do plants like trees survive freezing temperatures? To understand that we need to zoom in to see the inner workings of the plant.

Like other organisms, plants are made up of cells which contain fluids, and so these cells can freeze when it's cold enough outside. When ice forms in a cell, it damages the cell (we experience this if we get frostbite). If enough cells are damaged, the organism won't survive. The wood frog we just mentioned avoids harm by using special antifreeze compounds to prevent freezing damage. Various species of fungi also use chemicals in their cells which act as antifreeze. Some plants, like pine trees, move water outside their cells—it sits between cells instead—and so when ice forms it doesn't do as much damage to the plant. This is one reason why coniferous trees (evergreens) like pines are often found in colder environments.

Just as some animals go into hibernation, most plants and fungi enter into resting stages too, which we call **dormancy.** Many seeds remain dormant throughout the winter, emerging in the spring as the ground thaws and temperatures increase—giving us beautiful spring wildflowers!

On the opposite end of the temperature scale, some desert plants have thin leaves to prevent damage from the strong sun. Others have spines (a type of modified leaf) which are white or fuzzy (like some cactuses such as the teddy-bear cholla). These spines reflect sunlight away from the plant, preventing the energy in the sunlight from being absorbed. This effect is similar to how we can experience a difference in temperature based on our clothing color—wearing black clothing in summer absorbs more sunlight and could cause us to feel warmer.

Like animals, plants can also use evaporation as a cooling mechanism. Yes, in a way, plants sweat! They release water through small pores in the leaf tissue called stomata; the evaporation of this water cools the plant during periods of high temperatures.

Temperature isn't the only environmental factor organisms need to deal with, though. Opening stomata and using water for evaporative cooling only works if you have enough water to begin with—this may be an option in wet places, but may not be an option in the desert. In the next chapter, we will look at another environmental factor organisms must deal with: the availability or scarcity of water.

FOUNDATIONS REVIEW

✓ When we study nature, we see that organisms are adapted to the environments and habitats they live in—that is, they have traits that match the challenges around them. An adaptation is a trait that helps an organism survive in its environment.

✓ Animal adaptations to temperature extremes include behaviors, like migrating south for the winter or hunting for food at night when it is less hot, or physiology, which is the makeup of their bodies and the functions of their parts and systems. An example of a physiological adaptation is sweating when it is hot.

✓ Even though plants can't do certain things animals can to deal with extreme temperatures (like migrating), they still have ways to survive. For example, they can enter a dormant phase during winter to save up the energy they need to grow again in the spring, and plants can use evaporation as a cooling mechanism (like sweating). They release water through small pores in the leaf tissue called stomata; the evaporation of this water cools the plant during periods of high temperatures.

The Desert Fathers

In this chapter, we saw how some organisms adapt to hot environments. In the next, we discuss water scarcity. Both of these challenges exist at the same time in desert ecosystems. There are many remarkable organisms that have unique traits which help them survive this extreme environment—scorpions (and the long-eared bats that eat them), ironclad beetles, sand cats, fennec foxes, and many more! Some human cultures have also learned to live in dry, desert environments. A few saints and hermits from centuries past sought solitude and prayer by moving into the desert.

The Desert Fathers were early Christians who sought to follow Christ by taking up lives of poverty, prayer, and work. In the early third century, they moved out into caves and other places within the deserts of Egypt. Life there was difficult. They had no air conditioning, pools, or cold drinks! Nonetheless, they persisted and offered up sacrifices to God. Perhaps they even learned some lessons from His creatures on how to persist in such hard conditions.

One of the first of these Desert Fathers was Saint Paul of Thebes. He fled into the desert when Christians were being persecuted by the Roman Empire. Even though he lived in the desert as one of the first Christian hermits, with all the hardships that life brought with it, tradition holds that he lived to the age of 113! One disciple of Saint Paul was Saint Anthony of Egypt (or Anthony the Great), considered by many to be the father of monastic living. He wrote out guidelines (a book of observances) to help instruct men who sought to live as monks.

Saints Paul and Anthony, pray for us!

Various groups of animals gather at a watering hole in South Africa to stay hydrated. No challenge is arguably more important to organisms' ability to survive than maintaining their water balance.

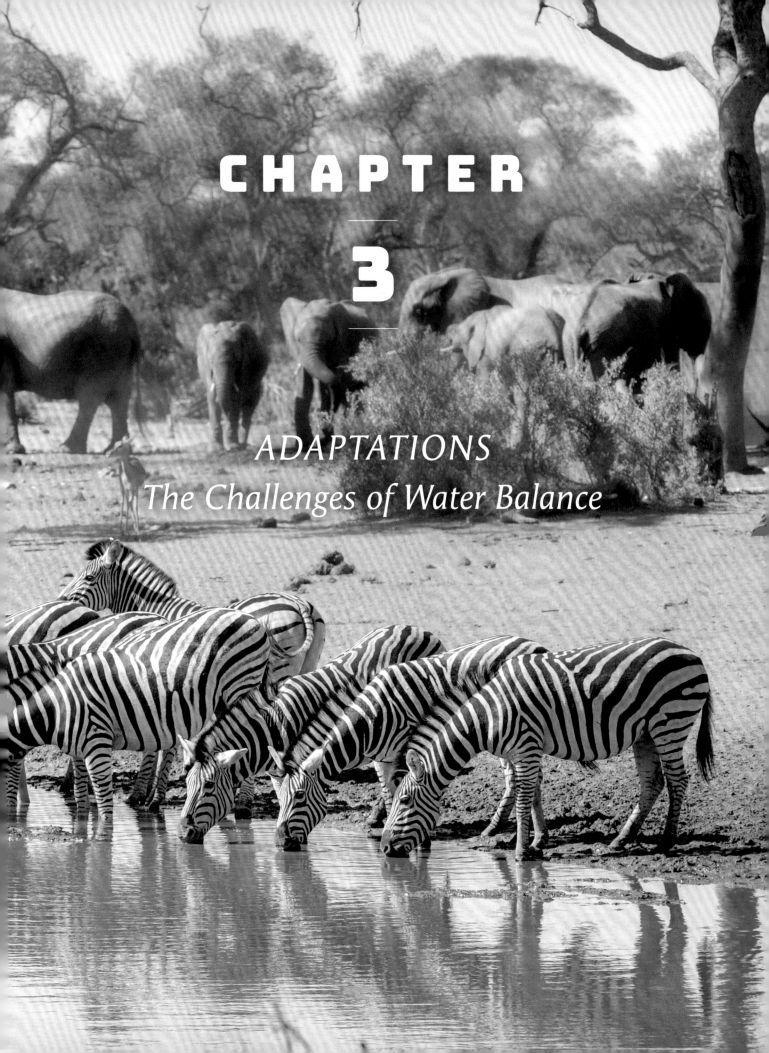

CHAPTER 3

ADAPTATIONS
The Challenges of Water Balance

THE NECESSITY OF WATER

The world is full of challenges for living organisms. We already read in the last chapter how some organisms manage to survive extreme temperatures. But temperature is not the only abiotic factor that affects living things. In this book, we will see a few more examples, focusing primarily in this chapter on the abiotic necessity of water (all living things need water to survive).

Knowing how important water is for all organisms, it makes sense that the availability of it in a biome would affect organisms that live there. The cells and bodies of organisms—whether plants, animals, fungi, or other living things—are made up of approximately 60 to 90 percent water, the exact figure differs among the species. With the greater portion of each organism consisting of water, it makes sense that the normal physiological functions of organisms depend on that water level. Organisms need to maintain what we call a **water balance**, taking in as much water as they lose to the environment in order to keep their interior water level consistent.

In the last chapter, we reviewed several ways organisms can use water to cool themselves off. Water is used up in bodily functions of organisms in this and other ways. Animals use water to process waste (through urination and defecation), and plants use water to conduct photosynthesis. This water loss must be replaced to maintain an overall balance in the "pool" of water the organism needs.

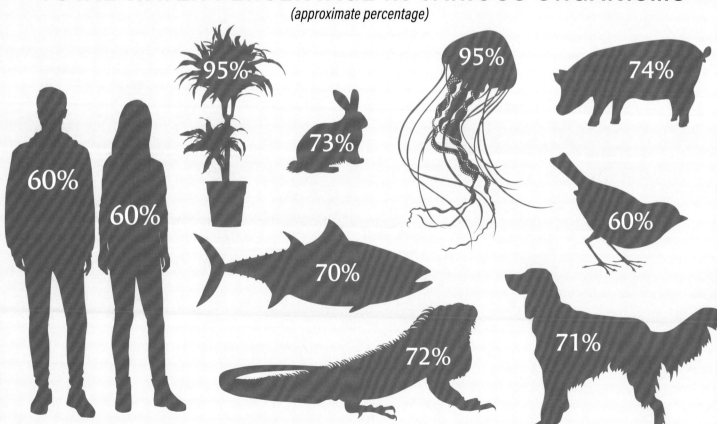

TOTAL WATER PERCENTAGE IN VARIOUS ORGANISMS
(approximate percentage)

In your own case, imagine your parents ask you to help them out in the yard by cutting the grass, picking up sticks, or raking leaves. Let's say it's a very hot, sunny day. As a way to cool, your body is going to sweat, which is a loss of some of your water. As a result, you'll have to go inside and drink water to even out your water balance or, as you might say in more familiar terms, to stay hydrated.

But if an environment has very little water available—through precipitation, or in the soil, or in the atmosphere—then it would be hard for an organism living there to maintain its water balance. It would be hard to stay hydrated after you do yardwork with your parents if the water main broke and the water was out. You would remain thirsty, and so you would have to find water somehow (a friend's house, bottled water from the store, etc.). Other organisms also need to maintain their water balance, but do not have the convenience of indoor plumbing or bottled water.

So what do plants and animals do to maintain their water balance in places where water is scarce? Let's take a look at some examples.

THE SEARCH FOR WATER

Deserts are a classic example of a place where water is lacking. Depictions of deserts in books and media might include someone traveling across a sunbaked landscape, thirstily drinking the last drops from their canteen. But other places outside of deserts experience a distinct lack of precipitation too, like the tundra. This lack of water might be year-round or it might be seasonal, meaning the precipitation amount depends on what season it is. One biome is called the tropical dry forest, which has dramatic swings from its rainy to its dry season. Up to seventy-five inches of rain can fall during the rainy season and as little as two inches in the dry season!

Miombo Forest in Malawi (left) and the Namib Desert in South Africa (right) are two of the driest places on Earth. Organisms that live in habitats like this may struggle to find enough water to survive.

So if all organisms need water to survive, how do they manage to thrive in these dry (or periodically dry) places?

Earlier we mentioned how plants have stomata, little pores where water is released. The stomata also let *in* carbon dioxide, which is critical for photosynthesis. The entry or exit of molecules through the stomata is controlled by two guard cells, one on either side of the pore, or hole (think of them like knights guarding the castle gate). Some desert plants only open stomata at night, which provides the carbon dioxide needed for photosynthesis without losing as much water to evaporation (which would occur more in the heat of the day). Many desert plants also have water storage capabilities—cactuses, for example, and other succulent plants store water in fleshy leaves or stems. These traits allow plants to store up water whenever it falls, and then hold onto the water they already have.

But plants also have to collect that water to begin with, right? Storing water doesn't do you much good if you don't have any water to store.

Most water collection by plants takes place in the roots, especially when the soil is moist. But what if the soil is arid (meaning dry) and no rainfall is expected for months? Some plants face these exact conditions. Much of the water available in a terrestrial environment is from rainfall or other precipitation that's absorbed by the soil. But the atmosphere also has some water— we feel this as **humidity**. **Dew** also forms in early mornings, which is condensation of the water in the air onto the ground and other surfaces. Some plants are well-adapted to dry conditions because they can gather dew or water from the air. The Namib Desert is an extremely dry region along Africa's southwest coast. Because of its proximity to the coast, though, fog rolls over the area often

In places like the Namib Desert early morning mist and dew are some of the only opportunities plants and animals have to soak in water.

enough for plants, and some animals, to use it for their water source. As water condenses from the fog onto grasses or other plants, it is channeled to drip down toward the base of the plant, where the roots can absorb it.

Believe it or not, animals can take in water from dew and fog, too. An estimated fifty species or more collect water from the fog in the Namib Desert. One beetle species (*Lepidochora kahani*, also called the flying saucer trench beetle) acts as a little architect, building dew collection traps out of trenches dug into the sand. In the United States, the Texas horned lizard (*Phrynosoma cornutum*) has a specialized set of tiny channels or grooves covering its scales. When the lizard positions its body in the right way (head down), gravity and these channels direct the dew or other moisture on the lizard toward its mouth so that it can take a drink. Remarkable!

You can see where this flying saucer trench beetle and Texas horned lizard get their names from. Both are creative in how they take advantage of dew and fog to help maintain their water balance.

Once an animal has collected some water, it can help prevent thirst by keeping as much of that water as possible. We find traits that minimize water loss in animals, just like we do in plants. For example, some desert animals have extremely efficient excretion systems (via differences in structures in their kidneys), meaning their urine is highly concentrated, having lots of waste output but using as little water as possible. Various desert mammals have fewer sweat glands than their non-desert counterparts—meaning they lose less water via evaporation, though this also means less cooling via sweat (instead, they must find other ways to cool off, like being active at night).

Another way to survive water scarcity is simply to skip over those times of year when water is hard to find. From the last chapter, we know hibernation is used to avoid difficult winter months that are too cold with too little food. Hibernation is also used to avoid other difficult environmental conditions, like water scarcity. The aptly named fat-tailed dwarf lemur stores fat in its tail to use as an energy source when it hibernates. Rather than avoiding cold, though, the lemur hibernates during the dry season in its tropical dry forest habitat. Plants in this forest also rest during the dry period, producing flowers and fruits when water is more available in the rainy season. Can you guess what the fat-tailed lemur eats? If you guessed flowers and fruits, you were right!

FAT-TAILED LEMUR

THE DANGERS OF TOO MUCH WATER

The examples of animals and plants and the abiotic necessity of water, so far, have all been about dealing with a lack of water. But too much water can also be a bad thing. Plants need water, as we have seen, but too much water can prevent roots from getting oxygen. Then the roots die and begin to rot, which spells disaster for the rest of the plant. It's almost like the plant is drowning. Because of this, only certain plants thrive in wetlands and areas that undergo persistent or periodic flooding. Bald cypress (*Taxodium distichum*) is a type of coniferous tree that produces knobby "knee" structures from their roots that emerge from the ground. Scientists think this may help their roots continue to receive oxygen during flooding, and also may help stabilize the plant in the soft, water-logged soil. Animals, too, have ways of avoiding seasonally flooded areas. For example, jaguars can keep to the treetops when the Amazon River is flooded each year.

Plants and animals both have ways of dealing with the presence of too much water. Bald cypress trees have roots that curl back up into the air to receive oxygen, and wild cats like this jaguar can climb trees to avoid floods.

FROM ABIOTIC TO BIOTIC CHALLENGES

In the past two chapters, we've learned how organisms have traits that help them survive in the biomes that they call home. We looked at various examples of plants and animals that can survive temperature extremes (hot and cold) and dramatically different water availability (water scarcity versus flooding). Though temperature and water availability are perhaps the two most important abiotic factors which affect organisms, they certainly are not the only ones. Wind, wildfires, soil content, salt balance, oxygen levels, and light levels are additional factors that can affect organisms.

But there are millions of species on Earth, and each has its own specialized features, so we really have only scratched the surface of unique and amazing adaptations. On your own, take a look at your favorite animal or plant, or maybe a few organisms from your favorite biome to see if you can identify features which help them survive in the face of all the many abiotic challenges present in their habitat. As we move forward in this book, we will also start looking at the **biotic** factors (living things) which create challenges for organisms—like predators and competition among species.

Ecosystems Fun Fact: The Atacama Desert in Chile (South America) is the driest place on Earth outside of the poles. It receives less than 1 mm of rain per year, and in some places has not received a drop of rain in 500 years!

FOUNDATIONS REVIEW

✓ The cells and bodies of organisms—whether plants, animals, fungi, or other living things—are made up of approximately 60 to 90 percent water. With the greater portion of each organism consisting of water, it makes sense that the normal physiological functions of organisms depend on that water level. Organisms need to maintain what we call a water balance, taking in as much water as they lose to the environment in order to keep their interior water level consistent.

✓ Plants and animals have to find water and store it, which is difficult to do in regions where water is scarce, like deserts. But organisms have unique ways of capturing the water they need, such as taking in water from fog or dew, or by entering a kind of hibernation or dormant state during dry seasons.

✓ While the most difficult thing is a lack of water, too much water can be a bad thing too. In plants, too much water can prevent roots from getting oxygen, causing the roots to die and begin to rot, which spells disaster for the plant. But certain plants thrive in wetlands and areas that undergo persistent or periodic flooding, like the bald cypress, which is a type of coniferous tree that produces knobby "knee" structures from its roots that emerge from the ground. Scientists think this may help the roots continue to receive oxygen during flooding, and also may help stabilize the plant in the soft, water-logged soil.

The Harmony Between Faith and Science

"God created the world to show forth and communicate his glory. That his creatures should share in his truth, goodness and beauty—this is the glory for which God created them." CCC 319

The last two chapters have discussed how organisms have unique characteristics specially adapted for their environments. These adaptations allow them to survive even in extreme conditions, like in the hottest deserts or the coldest tundra.

But the world around organisms is not unchanging—temperatures, rainfall, predators in the area, and available shelter can all fluctuate. In the next few chapters, we will investigate how mechanisms like natural selection can alter the adaptations we see in nature. These changes allow populations of organisms to adjust to their surrounding environments. The ideas of adaptation and change within populations are central to our understanding of the many aspects of biology, and so are important to be included in our book on ecosystems, populations, and organisms.

The shifting and changing of populations of organisms we will explore is, by technical definition, an evolutionary change (though perhaps a small one). Indeed, some may already be aware of the connection between terms like adaptation and change, natural selection and evolution.

In the following two chapters, we will generally highlight ways that populations of animals and plants have changed over time in *small* ways. When natural selection is involved, we will see that these changes often result in a population that is better equipped to deal with the environmental challenges it faces. These small changes over generations is a process that is often referred to as microevolution (*micro* = small or minor). For example, the average size or shape of the beak of a population of birds may change as the food source changes, or bacteria populations may develop resistance to our antibiotics, or a plant species may become more suited to growing in extreme temperatures.

Our book's focus is on the processes of adaptation and natural selection (as a form of microevolution), because it directly applies to our discussions on the relationships between organisms and their environment. As you read through, you may be aware that there exists a broader theory of evolution which includes changes on a larger-scale. This broader theory asserts that smaller changes add up over millennia to become much larger changes (including generating new species and the diversity of life we observe in nature). While discussions on the science behind this larger set of evolutionary theory, and how these ideas relate to our faith, are important, it is simply not the focus of this book. We do not have the time or space to fully explore these ideas, and so we will limit our focus to *microevolution* and the observable processes of adaptation and natural selection.

As Catholics, we know through faith that God created and sustains life on our planet, just as He watches over our lives. The Catholic Church also knows that faith and reason cannot be in conflict. Thus, we are invited to carefully evaluate scientific evidence of how life has adapted and changed over time, just as we would do with other theories in science—and to do so in the light of reason. When you do have the chance to study evolutionary theory more in-depth alongside your parents and teachers, my hope is that you do so confident in your faith, knowing truth cannot contradict truth. In the meantime, this book will explore some of the fascinating changes we can observe in nature, while we also together reflect with awe at the Author of the universe, wherein all these wonders exist!

A flock of butterflies take flight from a large boulder wedged in the midst of a steam.

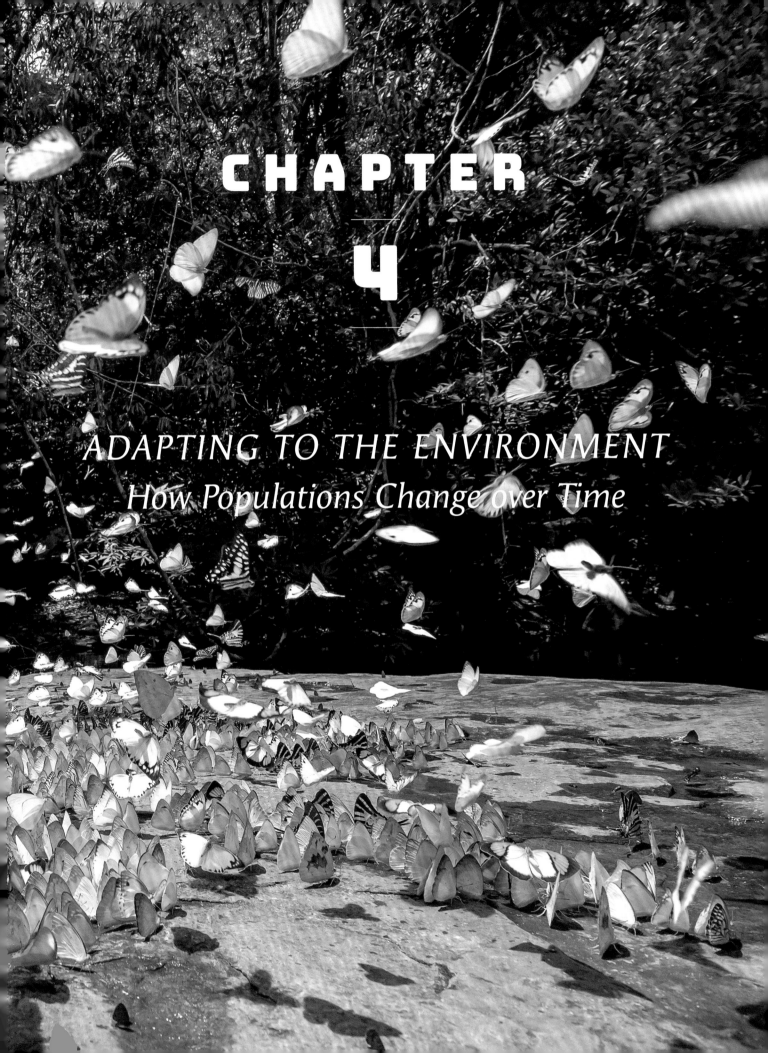

CHAPTER 4

ADAPTING TO THE ENVIRONMENT
How Populations Change over Time

THE STRUGGLE TO SURVIVE

In the first chapter, we learned that ecology is the study of many different levels of organisms: individuals, populations, ecosystems, and more. This chapter focuses on natural selection and the process of adaptation, two important topics in ecology and in all of biology. These processes result in changes in populations, often such that future generations are better fit to overcome challenges in the environment. In biology these changes are defined as evolutionary changes because, at its most basic definition, **evolution** is a change in the characteristics (or traits) of a population of organisms over time, or over generations. As a reminder, our textbook limits discussions to the "micro" changes to populations of organisms, via processes like natural selection. In the next two chapters, we will see effects on two levels of ecological organization: how *populations* can change over time (evolve) by the success or failure of *individuals* within that population.

PEA PLANT

To explore how a population of organisms could change over time, let's use a simple example. Organisms have physical characteristics and traits—pea plants, for example, can have purple or white flowers (a physical trait), and round or wrinkled seeds (another physical trait). Just to clarify, these physical traits we see are determined by the genetics of the organism written in its cells, meaning it has genetic "code" which gives it either one color or the other, not both (more on genetics in the next chapter). This is like how dogs, cats, or other pets have a particular coat color (a physical trait, which would be connected to their genes), but they don't change it each day—it's just the way they are. Sometimes the traits an organism has can determine its success. For example, if you let a population of pea plants naturally grow and reproduce in a wild field, the mix of different flower colors within the population might change over time.

How could that be?

Well, let's say when you planted the first seeds, only 10 percent of the plants (one out of ten) had white flowers. But now let's assume that over the next five years, bees or some other pollinator preferred pollinating white-flowered plants. In this hypothetical garden, white-flowered plants would be more successful because they are pollinated more frequently, resulting in more white-flowered pea plants being produced. This means that at the end of five years, the white-flowered plants would probably be more plentiful—maybe 70 percent more.

The exact number isn't important. What's important is that there was *an increase* in the proportion of white-flower individuals in *the population* (in the next chapter we'll also discuss the genetics underlying changes like this, because the DNA or genes that code for white flowers also increased in the population). Individual plants succeeded more often with white flowers because they were pollinated more frequently, leading to more white-flowered individuals in the population in future generations.

Adaptive Processes versus Adaptive Traits

Adaptation can, perhaps confusingly, refer to two different things. It can be used to describe a process of change over time, but also physical traits or characteristics an organism might have because of those processes which help it survive in its environment. So an adaptation (a trait) can be the result of adaptation (the process). To help resolve this confusion, we could also call traits that result from the process of adaptation **adaptive traits**. Over many generations in cities, for example, populations of ants have become more tolerant of heat. Because cities have lots of hot asphalt and other surfaces which heat up quickly, ants that were the most heat tolerant were the most successful from generation to generation. *Adaptation* (as a process) describes the changes in the ants' physiology over time, which lead to the *adaptive trait* of heat tolerance.

Through simple processes much like this example, we will see how populations can sometimes adapt to new and changing environments—for example, if new bees that prefer to pollinate purple flowers show up, the population on average shifts to result in more purple-flowered plants. We defined adaptation in chapter two, but let's remind ourselves again: In biology, the process of **adaptation** describes the ways in which organisms become more suited to their environment.

The last few chapters discussed how organisms have special and sometimes unique traits to help them deal with challenges in their environment—like animals using fat to keep warm, or waxy surfaces on plants to preserve water. Generally speaking, we expect organisms to be more successful in their naturally occurring habitats. For example, a desert-dwelling rodent would have better success in a hot, dry climate than a rodent from a cool, wet forest. The desert-dwelling rodent has traits which help it survive high temperatures and the scarcity of water (for example, reducing water lost through sweat or waste, or being active at night).

But how did that rodent adapt to the environmental challenges of the desert in the first place? Can organisms change and respond to new challenges over time? It turns out, yes! Now, it's possible forest rodents would need too

The greater Egyptian jerboa (a mouse-like rodent) has long legs to help it hop like a kangaroo through scorching hot terrain, while the beaver's thick coat of fur keeps it warm in frigid temperatures. Both of these traits are adaptations that help them survive in their environment

many new and different traits to be able to thrive in a desert if we transported them there today. But we can easily see subtler changes *within* species throughout nature. One good way to examine the process of change over time is to look at a single species that lives in multiple environments—like a species of mouse that could live in either urban or rural places. Examining the differences between the individuals in a given population in each environment can demonstrate which adaptive traits may be important in each place. City mice may need different traits to succeed, and so we see differences between the city mice and the country mice (like with the ants in the call out box on the previous page).

To better understand this, let's take a look at an example with plant species living at various elevations—each elevation will have different environmental conditions, which in turn affect the individual plants living there.

LOCAL ADAPTATION IN PLANTS

A single plant species may have populations living at both high and low elevations, such as up on a mountain and down in the valleys below. The high elevation populations would experience colder temperatures, and so may need slightly different traits than their relatives living at low elevations. But what if you took plants from the low elevation and planted them up on the mountain? They might still survive but would likely not thrive quite as well. Plants transplanted to a new elevation may grow more slowly, compete less well against other plants, produce fewer seeds, or just otherwise exhibit struggles to survive. This would show that the low elevation plants have traits which help them in their valley habitat (in warmer temperatures, perhaps with more moisture in the soil), whereas the high elevation plants have traits which help them on the mountain (where it may be colder, as we said). In other words, *within the species*, each population has adaptations to its particular home. This is what we call **local adaptation.** But how does this come about?

The edelweiss is a rare wild flower that can grow high up on a mountaintop. It has local adaptations that help it survive in cooler, thin air.

Insects, like these swarms of mosquitoes, are just one of many factors in an environment that can lead to organisms developing adaptations. Insects can be a pest to animals (including us!) and can be deadly in some cases to plants.

It turns out populations can slowly adapt to new environments over time, with the right tools to do so. Let's imagine that our plant species started out at low elevations, but somehow it expanded up the mountainside; perhaps a storm carried seeds up, or a curious scientist moved them to study adaptation. If there were certain traits which already existed in the low elevation plants that were helpful at high elevations, then plants with those traits could survive and produce more seeds—passing those successful traits on to their offspring. Over time, those successful traits would increase in the population, and any harmful traits would decrease in the population. In effect, living in the new high elevation environment would "weed out" negative traits.

We can actually measure the effects of selection on plant traits based on elevation. For example, within a species, individual plants tend to have smaller, thicker leaves at higher elevations—this leaf shape is better for colder temperatures at higher elevations. What happens to give us this repeated pattern across plant species? Some individual plants happened to have smaller leaves, while others had larger leaves. As the population spreads out, moving up the mountainside, the plants with smaller leaves would arrive already better prepared to deal with the cold mountaintop habitat. Over time, the proportion of plants with small leaves would increase in the mountaintop population. This means that in the end, the mountain population would have smaller leaves, on average, than the valley population.

EXAMPLES IN THE ANIMAL KINGDOM

In this way, there can be some flexibility within a species in adapting to various environmental challenges. For example, in more arid or dry environments, mosquito populations can regulate water balance better, and so have a higher resistance to drying out—in part by having a larger body size. If variation in body size exists in the mosquito population, and if larger mosquitoes tend to survive best in dry environments, then eventually we would expect populations of mosquitoes in dry environments to have a larger body size, on average.

Adaptation Fun Fact: Some vegetables are better than others when it comes to growing at high elevations. So if you live in the mountains, make sure to spend time in your garden growing beets, carrots, lettuce and potatoes.

This is because the small ones tend to die sooner in these dry places and have fewer offspring. In the end, through the success and failure of individuals with certain traits, the mosquito population is better adapted to the dry environment.

As another example, could an animal or plant species get better at dealing with hot or cold environments—those abiotic challenges we explored in chapter 2? Yes! Some snail populations show variation in shell color—from lighter yellow to darker brown. When the snails move into areas with hotter temperatures and less shade, or as climates warm, the darker shells absorb more heat, which dries up their water supply. Thus, individuals with dark shells don't live as long and have fewer offspring in the new hotter environments, which results in mostly yellow-shelled populations. The population, as a whole, is now better adapted to the hotter environment.

All these examples are similar to the one we used when we began our chapter about the purple and white pea plants and the bees pollinating only the white-flowered plants. In these examples, the populations change over time—they evolve—when the average traits in the population change. This often happens because some traits are better than others for the current challenges the population is facing. Importantly, the environment dictates which traits are best to help an organism survive. If the environment changes back, the population could respond to the new challenges. So if bees pollinated purple flowers instead, our pea populations would shift back toward purple; or if the climate cooled, darker-brown snail shells might be advantageous again.

ADAPTATION VERSUS RANDOM CHANGES

So far we have looked at examples of how populations change when faced with challenges in their environment. When these challenges arise, then changes in the population lead to more or better adaptations—the population is better fit to the environmental challenges it encounters. We refer to these specifically as

Flooding is an example of an abiotic factor that leads to genetic drift. When populations dramatically decrease due to catastrophe like this, the remaining surviving population looks different than the original.

adaptations. However, changes can occur in a population over time for other reasons as well.

Sometimes populations change in random or negative ways. For example, let's go back to our pea plant garden with purple and white flowers. If the bees that prefer white flowers are still visiting the field, pollinating all of the white-flowered plants, then white-flowered plants still have an advantage. But what if something random or chaotic happened in the field? Say a tree fell on it, or part of it flooded during a dramatic rainstorm. Some pea plants would die off, and this could potentially change the traits of the population as a whole simply because only a certain subset would remain. Maybe, by random chance, white-flowers went from 70 percent of the population back down to 20 percent after a flood. This still technically counts as a change in the population over time, but it results in a population that is less well equipped to be pollinated by the bee population. The flood-induced change in the population is an example of genetic drift, or a population bottleneck. **Genetic drift** describes random fluctuations in the traits of a population over time, and it is one type of change that does not necessarily lead to better adapted populations.

This chapter focused on using examples to help introduce us to the concepts of adaptation and change in populations. Evolutionary changes occur as the average traits of a population change over time. Often these changes occur in an adaptive way, causing populations to be better "fit" to their environment (we saw a brief example where this was not the case with genetic drift). Hopefully these examples will help as we move forward with this same topic in the next chapter, where we will provide more specifics and some terminology to help us understand these important biological concepts.

Remember:
An adaptation (a trait) can be the result of adaptation (the process).

FOUNDATIONS REVIEW

- ✓ Populations of animals or plants or other organisms evolve over time whenever the traits of that population change. Because successful traits are passed on more often than negative ones, the population can become better adapted to its environment over time. These changes can then help organisms survive in all sorts of environments.

- ✓ Within the same species, each population can have adaptations to its particular home. This is what we call local adaptation. It refers to different populations within the same species developing different traits as they adapt to different places or environments.

- ✓ The changes in environment that bring about these adaptations can depend on the temperature, the amount of water or sunlight, predators in the area, or any number of other factors (also called "selective pressures") that could influence an organism's survival, including random events like a flood or a fire.

The Father of Modern Genetics

In this chapter, we discussed pea plants as a way to introduce the topic of changes in populations. This is also one of the most classic examples used to demonstrate how genes connect to physical traits and how traits are inherited from parents to offspring. Such knowledge comes to us at least in part from the careful experiments of Gregor Johann Mendel, OSA (Order of Saint Augustine). This Augustinian friar, who also happened to be a brilliant scientist, used peas to first demonstrate many of the things we know today about genetics and heredity (or, how traits are inherited—an important part of natural selection—we'll explore heredity in the next chapter!). Mendel documented all sorts of pea plant traits (plant height, pod shape and color, seed shape and color, and flower position and color) and tracked when and in what circumstances young pea plants looked like their "parents." He tended his garden and carefully studied peas for more than seven years, during which time he grew and observed almost thirty thousand plants!

Gregor was born Johann Mendel in 1822 in the Austrian Empire, in a place that is now within the Czech Republic. He was born into a farming family, and grew up in agricultural areas where he learned about growing crops, working in orchards, beekeeping, and living off the land. During his life he was also interested in other scientific disciplines besides biology, such as meteorology and mathematics. He studied some of these topics at university before joining the Order of Saint Augustine, taking the name Gregor. He studied to be a school teacher but failed his teaching exams more than once. After failing an exam in 1856, he instead sought permission to do research in the gardens of St. Thomas' Abbey in the city of Brno. That research would lead to his discoveries about pea plants and genetics.

Interestingly, the importance of his work with pea plants was not realized until several decades later when several scientists replicated his work, thereby verifying it and publishing the results. Scientific studies are often replicated—repeated again—to show that the results could not be due to chance, but rather are a pattern found in nature.

Gregor's life provides a good lesson in perseverance. Though he was probably disappointed when he couldn't become a school teacher, God clearly had other plans for him. Gregor Mendel didn't discover everything we know about genetics, but for his work he is now known as "the father of modern genetics." His story shows us that science builds upon itself over time so that we can better understand God's creations.

This great gray owl's feathers disguise it against the backdrop of a tree in the forests of Canada. Such an adaptation allows it to hide from the prey it stalks.

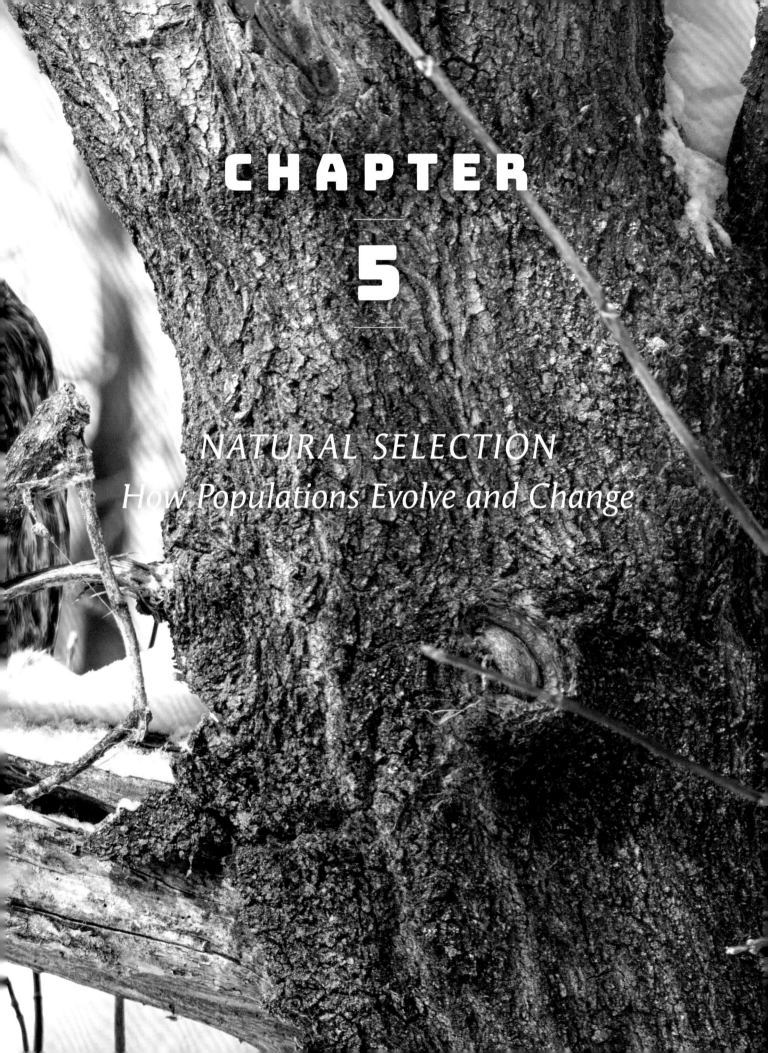

CHAPTER 5

NATURAL SELECTION
How Populations Evolve and Change

FITNESS, DNA AND GENES, AND HERITABILITY

We will continue our discussion of evolution in this chapter by examining what scientists call the process of natural selection. The simple adaptation examples from the previous chapter described the process of **natural selection.** Here we will see how these examples fit within the formal definition, which is the process by which organisms better adapted to their environment, or those that are a better "match" to their environmental challenges, will produce more offspring.

But before we begin, there are a few important biological concepts that we will briefly discuss. These are (1) fitness, (2) DNA and genes, and (3) heritability. Some of these terms may be more familiar than others, but they are all important for us to know as we seek to understand the process of natural selection and its connection to changes in populations of organisms.

Fitness: When we say a plant or other organism "succeeds" or "fails" compared to others, this describes **biological fitness.** You have probably heard the word fitness used in relation to human beings before—it describes how well our bodies are conditioned and trained to deal with physical activity. Biological fitness in ecology (as in other biological sciences) has a similar meaning in that it is related to how well prepared an organism is to deal with the challenges it faces. Biological fitness doesn't just measure physical ability, it measures the reproductive success of an animal, plant, fungus, or other organism.

Fitness is described by many factors, but we can summarize them by saying that the organism needs to survive within its environment long enough to reproduce. So traits (whether physical or behavioral characteristics) that *increase* survival would *increase* fitness, and traits that lead to more offspring would also

Though it may be hard to believe, these giant oak trees started out as acorns!

increase fitness. As an example, an oak tree that produces more acorns (which then turn into more oak trees) has higher fitness than a neighboring oak tree that produces fewer acorns. Any number of factors could help it produce more acorns—it grows larger or taller, it has more access to light or nutrients, or it lives longer—and so all of these things affect the oak tree's fitness.

DNA & Genes: The traits of an organism are determined, at least in part, by its genes. **Genes** are segments of DNA (deoxyribonucleic acid) which contain the instructions for how to build the various proteins, or develop the parts that

make up an organism. All life on Earth has the instructions on how to build it "written" in the DNA contained in its cells. In this way, we say that DNA is an information-carrying molecule. Each of our own cells—skin cells, lung cells, liver cells, etc.—has our entire set of DNA instructions within it (you can learn a lot more about DNA and cells in the *Foundations of Science* book called *Cells & Systems*).

Heritability: DNA also transfers this important information from one generation to the next. For example, we know that pea plants might have purple or white flowers, and these physical differences are determined by the code written in their DNA. That is, each plant has the "instructions" for growing either white or purple flowers. Pea plants with white flowers develop seeds (the next generation of plants) with DNA instruc-

BORDER COLLIE

tions on how to develop white flowers—the "parents" pass this information on to the next generation of pea plants. If purple-flowered plants are more likely to have purple-flowered offspring, and white-flowered plants are more likely to have offspring with white flowers, then we say this trait is **heritable.** This should sound like the word "inherit," which you may be familiar with. We inherit traits from our parents too; that is, we receive certain physical characteristics from them, like brown hair or being tall. This is why we probably look much like our parents and grandparents, and likewise, why pea plants of one generation look like the ones that came before them.

Let's go back to our example from the last chapter of the plants growing at various elevations, including up on the mountainside. We mentioned the adaptation of small leaves helping growth at high elevations, so we would say plants with

Partially Heritable Traits

Some traits are *partially* heritable—meaning they are determined only in part by genes. One example is human height. Tall parents are more likely to have taller children, but things like access to good nutrition also helps children grow tall. So children are not *exactly* as tall as their parents, but instead development, nutrition, and other factors play a role. One estimate suggests that height is 80 percent genetic—so overall, children are still *more likely* to be tall if their mom and dad are both tall.

smaller leaves have *higher fitness* at high elevations (smaller leaves perform better on cold mountaintops, helping the plants succeed). To get those small leaves, we would expect the leaf size to be regulated by a particular set of DNA instructions that make small leaves grow. In other words, plants with small leaves would have a slightly different "code" which results in smaller leaves developing. Finally, if plants with smaller leaves tend to have offspring with smaller leaves, we would say that leaf size is heritable (they will pass this trait on to the next generation of plants).

THREE CONDITIONS OF NATURAL SELECTION

So organisms have DNA which determines their traits, and these traits can influence fitness (their ability to thrive and survive). Evolutionary changes can happen when the average traits of a population change over time, favoring those traits which lead to higher fitness. In particular, natural selection requires three things.

1. **Variation (change) in the trait.** If there is no variation in the trait, then nothing can change in the next generation. Selection acts on the variation already existing in the population. For example, if we were studying eye color in a population of pigeons, but all pigeons had brown eyes, it wouldn't matter which parents were most successful in raising chicks—the next generation of pigeons would always retain the same eye color. All parents and all offspring, in this case, would have brown eyes, regardless of which pigeon parents hatched the most chicks. In contrast, imagine another population of pigeons in which half had brown eyes and half had orange eyes—variation (difference) in eye color now exists in the population, and selection could favor one of the eye-color types. If pigeons with orange eyes were more successful in raising chicks, then we would expect more than half of the next generation of pigeons would have orange eyes simply because more orange-eyed chicks would hatch.

2. **The trait is heritable.** Remember that this means that offspring will resemble their parents—they inherit the trait from their parents. The previous example with pigeons only works if our assumption that orange-eyed parents have orange-eyed chicks is true. Traits can be

Species Fun Fact:
Another trait of pigeons is that they can have four different wing patterns: bar, barless, check, or t-check.

affected by the environment or genetics (the DNA of the organism), or both factors can affect traits simultaneously. But if something is mostly determined by the environment, then the fitness of parents with particular genes or traits will not affect how many offspring in the next generation have that trait. If pigeons develop orange eyes when hatched in the warmer summer months (temperature is an environmental effect), not because of their parents' eye color, then it wouldn't matter which type of parent (orange or brown eyes) laid the egg—summertime chicks would have orange eyes regardless of their parents' eye-color.

3. **There is selection on the trait.** Another way to say this is that there is a connection between the trait and fitness—that is, having one trait is more likely to lead to survival and reproductive success. This selective process is what would slowly shift how many birds in our hypothetical pigeon population had orange eyes. Imagine if our population of pigeons were domesticated, meaning that humans were keeping them for food or for show or as pets. If the human keepers of the pigeons preferred orange eyes (as prettier or healthier birds, perhaps), they could increase the fitness of orange-eyed birds by allowing them to breed more often (or perhaps by more often using the brown-eyed pigeons for food). Over a few generations of **artificial selection** like this (humans intentionally engaging in artificial selection), the flock of pigeons could become entirely orange-eyed. Here the human owner deliberately introduced a selective pressure, causing higher fitness for orange-eyed birds.

If these three conditions are satisfied, then we can observe a change in the traits of a population because of the action, or pressure, of natural selection. In this particular pigeon eye-color example, we would technically call this artificial selection because human pigeon owners chose a particular trait to favor. Farmers often do this with agriculture and livestock, for example breeding cows that produce more milk, or corn that has larger ears with more kernels. Even more amazing, broccoli, cabbage, kale, cauliflower, brussels sprouts and more all originate from the same wild mustard plant. Agriculturalists select for the growth of particular plant parts, leading to dramatic differences in these vegetables. All the various dog breeds around the world are another great example of artificial selection—dogs can be bred to help with livestock, chase rodents, run quickly, look adorable, or any number of other traits.

So what happened in these examples of artificial selection? In scientific terms, we would say, for example, that dairy cows evolved over time

Pigeon Eye Color

This example of pigeon eye color is just a hypothetical example, but domestic pigeons do have three eye colors (dark brown, orange, or pearl/white). Could this eye color affect fitness in the wild? It may seem unlikely that eye color would have an effect on the survival of the pigeons (unless it was connected to visual ability in some way). But fitness is not only about survival. Birds often choose mates based on coloration, and if orange eyes were more preferred, orange-eyed individuals would more often find a mate, and so be more likely to make a nest and raise young (higher fitness!). We don't know that this is the case, but this is an example of a testable hypothesis about fitness. We could measure mate choice or nesting success of birds with different eye color to try to find an answer.

to produce more milk—the average traits of the population changed over time. Returning to the pigeons, once there is an increase in the proportion of birds with orange eyes in a flock, we would say that the pigeon population evolved via artificial selection. Changes in populations are often easier to visualize in artificial selection examples. But in a similar way, natural selection occurs everywhere in nature, not based on traits humans think are beneficial but based on whichever traits help the organism *survive and reproduce* in its environment. From the last chapter, selection favors smaller leaves at high elevation (producing larger populations of plants with smaller leaves than low-elevation populations), and yellowish-shells on snails in hotter climates (producing populations with more yellow shells than other populations).

With either natural or artificial selection, however, it is important to remember that the *population* evolves, or changes, over time. There is a change in the number of individuals within the population that have each trait, and this change occurs over generations. But individuals do not evolve. Each individual pigeon either has orange or brown eyes, and then either has higher (or lower) fitness compared to other pigeons. Depending on the success or failure of the individual pigeons in the population and the traits they pass on to their young, we see changes in the makeup of the population *over time*. The same is true for any natural selection example we study.

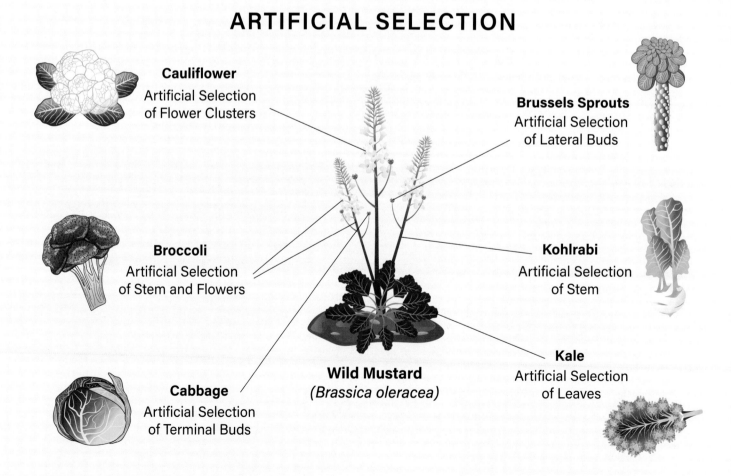

ARTIFICIAL SELECTION

Cauliflower — Artificial Selection of Flower Clusters

Brussels Sprouts — Artificial Selection of Lateral Buds

Broccoli — Artificial Selection of Stem and Flowers

Kohlrabi — Artificial Selection of Stem

Cabbage — Artificial Selection of Terminal Buds

Wild Mustard *(Brassica oleracea)*

Kale — Artificial Selection of Leaves

Another important point is that natural selection is not a planned or intentional process—populations or organisms aren't setting out to do something "for the good of the species." It is an orderly process, in that traits that are beneficial are predictably the ones that increase in a population. But unlike artificial selection, where we humans may have some end goal (more corn!), natural selection is not pursuing an end goal, and neither are the organisms involved. Rather, better traits persevere and worse ones are selected against, depending on how well they match the current environment. Moreover, the environment can change, which can change the target or direction of selection. If the bees pollinating the pea plants switched to only pollinating purple flowers, having white flowers would no longer be advantageous.

This closes our segment on natural selection, though we will see connections to it in other chapters, such as prey populations that respond to selection applied by the predators who hunt them. This process is woven throughout biology, as populations adapt to all the many aspects of their ecological environments that can change over time.

Remember:
Three conditions must be met for natural selection to occur: (1) There must be variation in the trait; (2) the trait must be heritable; and (3) there must be selection on the trait.

FOUNDATIONS REVIEW

✓ When we say a plant or other organism "succeeds" or "fails" compared to others, this describes biological fitness. This is a term that describes the reproductive success of an animal, plant, fungus, or other organism. Fitness is described by many factors, but we can summarize them by saying that the organism needs to survive within its environment long enough to reproduce.

✓ DNA molecules contain the instructions that give an organism certain characteristics and traits. These instructions can be heritable, meaning they can be passed from one generation to the next (though other environmental factors can also play a role in the traits and characteristics an organism has). Natural selection can occur whenever these three conditions occur: (1) a population contains variation in a trait, (2) that variation is heritable, and (3) the variation affects an individual's fitness (individuals with one version of the trait have higher fitness, on average, than others).

✓ Sometimes human beings engage in what is called artificial selection, which is when we pick and choose what traits we want a certain organism to pass down. For example, we might breed bigger cows that produce more milk for us, or bigger ears of corn that produce more kernels. This mimics the mechanisms of natural selection, which take place in nature based on what increases fitness (that is, what helps an organism survive and reproduce).

Silent Crickets and Playful Lizards

Some evolutionary changes take place over long periods of time, but we also have lots of neat examples of populations that change within generations or in just a few decades. For example, male crickets typically "sing" (chirp) to call out to female crickets, and this is how many species of crickets find their mates—including on the islands of Hawaii. But a parasitoid fly that lays its eggs on crickets was introduced to the islands some time ago, and the fly larvae eat the cricket from the inside. Not a good outcome for the cricket, to be sure—lower fitness if you're eaten up by fly larvae! But some male crickets carried a mutation, or a change in their genes, which gave them wings that cannot sing. These males had an advantage in the new environment with the parasitoid, since the parasites had difficulty finding them when they were silent. It may be more difficult for them to find mates without singing, but at least the flies can't find them! Over several years, scientists watched as the "silent" males increased in the population, to the point where some Hawaiian islands only have silent males.

Another neat example of a population changing over time is the side-blotched lizard, which plays a game of Rock-Paper-Scissors. If you've played this game, you know each symbol (rock, paper, and scissors) wins against one other type. In other words, no single one can outcompete all the others. But how do lizards play Rock-Paper-Scissors? After all, they don't have hands, do they? Well, their game looks a little different, with three male color types: orange, blue, and yellow. These physical colorations are connected to genes, to DNA that codes for each color type. Orange males are bullies and push around the blue ones, but yellow males look sort of like female lizards (females have yellow coloration) and so they can sneak past the bullies. In a population of mostly orange lizards, yellow has the advantage, and so yellow males pass on their genes more often and yellow increases in the population. But blue males are better at protecting against yellow "sneaker" males, as they cooperate to be vigilant against them. So in a population of yellow males, blue males and their cooperation strategy is best (because orange doesn't do well against yellow). Once blue males become common, orange males once again become the most successful (highest fitness) because they bully the blue males.

So the abundance of these three different types of males cycles up and down, in a kind of balance, because each color type has another that "defeats" it. Whenever one type becomes abundant, it "loses" to its color nemesis. These cyclic evolutionary changes in the population caused by the game of Orange-Blue-Yellow has been going on for millennia!

As a biologist, I am fascinated by all the amazing examples of evolution in nature, and how natural selection shapes populations over time. I am glad to be Catholic, knowing that we can study this wondrous, dynamic aspect of God's creation.

"As to the Divine Design, is it not an instance of incomprehensibly and infinitely marvelous Wisdom and Design to have given certain laws to matter millions of ages ago, which have surely and precisely worked out, in the long course of those ages, those effects which He from the first proposed."

–St. John Henry Newman in a personal letter, May 22, 1868

A crocodile surprises a herd of wildebeests drinking by the riverside. The presence (or absence) of apex predators in an environment can have drastic effects on the population of animals lower on the food chain.

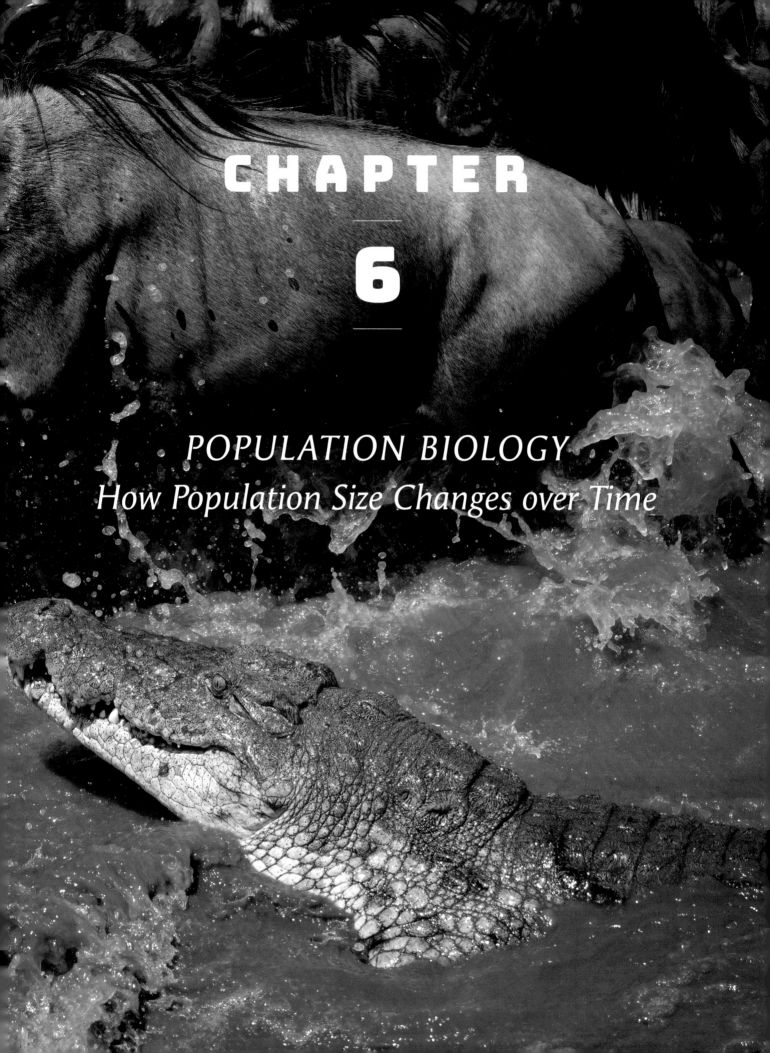

CHAPTER 6

POPULATION BIOLOGY
How Population Size Changes over Time

POPULATION EQUATIONS

One level in the study of ecology is the populations of organisms. The last two chapters showed how the attributes of a population could change over time, with advantageous traits becoming more common and disadvantageous ones becoming less common (or they could even disappear from the population). Part of that process of change involved which individuals lived longest or had the most offspring (that is, had the highest fitness). This chapter explores how the *size* of a population is also affected by survival, the birth of offspring, and death and other processes. As an area of study within ecology, we could call this the study of **population biology** (we discussed briefly something similar—population dynamics—in chapter 1).

To study populations, some mathematically-minded biologists might construct equations to describe the change in population size over time. I'm sure you all love math as much as I do, but we'll keep things simple for this chapter.

The terms in a population equation would include how many births, deaths, and immigrations and emigrations happen during a given period of time. These are the ways individuals enter or leave populations of organisms in nature. Births and deaths are likely more familiar terms to you. Songbirds lay eggs every spring, and nestlings hatch and fledge, adding to the population. But during that time, some adult birds (and also eggs or nestlings, unfortunately) may die from **predation** (one animal preying on another, from the word "predator") and disease—*subtracting* from the population. So births and deaths cause the population size to increase and decrease.

Song thrush hatchlings burrow down in their mother's nest. Building nests high off the ground helps protect their young from some predators, but not all!

The other two terms—immigration and emigration—describe the movement of individuals from place to place. **Immigration** simply means individuals joining the population from elsewhere, while **emigration** refers to individuals leaving the population to go somewhere else. These terms are used with human beings too. When an individual leaves his home country, he is emigrating, but when he arrives in his new country, he is considered an immigrant. These terms should also make you think of "migration," when animals move from one spot to the next.

But in the world of nature, where do individuals come from and where do they go? Remember a population is a distinct set of organisms living in a particular area. So we could study the black bear population in Great Smoky Mountains National Park, in Tennessee. Perhaps a black bear with a tendency to wander leaves its birthplace in Congaree National Park, in South Carolina, some 200 miles away. If it arrives at Great Smoky Mountains, then we would say it emigrated from the Congaree population, immigrating to Great Smoky Mountains. It's basically like he moved and changed addresses! The populations of black bears would change by one, with one fewer in Congaree and one more in Great Smoky Mountain.

A black bear forages in the forests of the Great Smoky Mountains.

The "Birth" of Organisms

Human biologists often use the term "births" to describe a growing population, perhaps because baby humans are born. But plants are not really born, and neither are insects, or many other organisms. Here in this chapter, we will use the term "birth" as shorthand, even though we may mean that a new individual is born or hatched or grown from a seed or spore. This is something to remember as we move through the material.

Population biology can help us predict whether endangered species are at risk of extinction—and maybe give ideas to help protect them!

Animals traveling alone like this bull moose (below left) can be more prone to danger than animals that travel in herds like these wildebeests (below right). But traveling in herds brings its own challenges, like fewer resources to go around (food and water) and the possibility of disease spreading quickly.

So populations increase or decrease due to births, deaths, immigrations, and emigrations. But as a population biologist, you might want to predict the changes that will happen *in the future*. Maybe you are working with **endangered species** (an organism that is in danger of going extinct), like orangutans or green turtles, and so you want to increase the population to keep them from going extinct. Or if you studied a destructive **invasive species** (any animal or plant not native to a given area that has a destructive effect), like the emerald ash borer (beetle), you would want to limit births and immigration. Let's examine the factors that influence population growth by affecting births, deaths, immigrations, and emigrations.

INFLUENCES ON POPULATION

One of the most obvious factors affecting population growth is the resources in the environment. For plants, this might mean water and nutrient availability. Predatory animals, like wolves, need prey items to eat, like deer or elk. But animals and plants also need resources like space. If you grow microorganisms in a vial in the lab, the population will grow quickly until the food and space start to run out within the vial. As the population grows, competition for food resources and space will increase. The same thing occurs in nature. A field already full of plants will have more competition for light, water, and nutrients.

Wolves living in an area with other wolf packs experience more competition for prey. There is simply less to go around. Imagine if you had one pizza to share with your brothers and sisters, versus one for each one of you. It's a bit like that.

A pack of wolves and a brown bear fight over a carcass. Competition for resources is what we call a density-dependent factor on populations.

Competition for resources is a good example of a **density-dependent factor** that affects populations. Density refers to how much of something is found in a given area. So the density of your room is higher if your whole family is in it versus if you are alone. Density dependent, then, means that it matters how many organisms fill up the space available. In a lab environment, many microorganisms densely packed into a vial experience more competition, whereas the same sized vial with only a few microorganisms would have less competition. Other density-dependent factors include disease and predation. For example, disease rates increase as the population increases. This is because individuals come into contact with other individuals more often as they compete for food and space, and this provides opportunities for bacteria and viruses to move from individual to individual and to spread more quickly through the population.

You can easily imagine that, as food supplies in an area decrease, mobile animals will leave to seek out new habitats. Food shortages or increased disease will also lead to more deaths. So competition and disease are density-dependent factors that lead to reductions in population size, through death and emigration. Moreover, these factors are strongest when the population is largest (more individuals means more competition and disease). This means that the population is most likely to crash when it gets too large. When there are lots of space and resources available (low competition), and disease transmission is low, more individuals will have sufficient resources to feed their offspring. New individuals may immigrate to the area, as well. This means that population growth is often higher at low (or intermediate) population densities.

Remember:

Immigration refers to new individuals coming in, while emigration refers to individuals moving out.

CARRYING CAPACITY

If the population decreases when it is large, and increases when it is small, we end up seeing a cycle of population size. Tracking population size over time on a graph often results in a wave-like pattern, with increases and decreases in population size as time moves forward. In a stable environment, the overall population often hovers near the **carrying capacity** for the population, even with these cyclical increases and decreases in population size. The carrying capacity of a population is exactly what it sounds like—the number of individuals, or the capacity, a particular habitat can "carry," or support. Let's look at a quick example on a graph.

A reasonable estimate of black bears in Great Smoky Mountains National Park is approximately 1,500 individuals. Therefore, that can serve as the carrying capacity for black bears in the park. If many cubs are born over a period of several years—let's say 500 more—but only a few bears die or emigrate—let's say 50—the population will exceed the carrying capacity by 450 because the population is now at 1,950 (1,500 plus 500 minus 50). In the years to follow, we would expect more deaths from disease, starvation, car accidents, et cetera, and more bears leaving the park to live elsewhere, sensing the crowded environment. Let's say 900 bears either died or emigrated. Now the population is at 1,050 (1,950 minus 900), or 450 below the carrying capacity of 1,500. The remaining bears would have more than enough food and space now (population size is less than the carrying capacity), and so it would be easier again to successfully raise cubs. In turn, the population would likely bounce back, and eventually exceed the carrying capacity again. And so the cycle repeats itself.

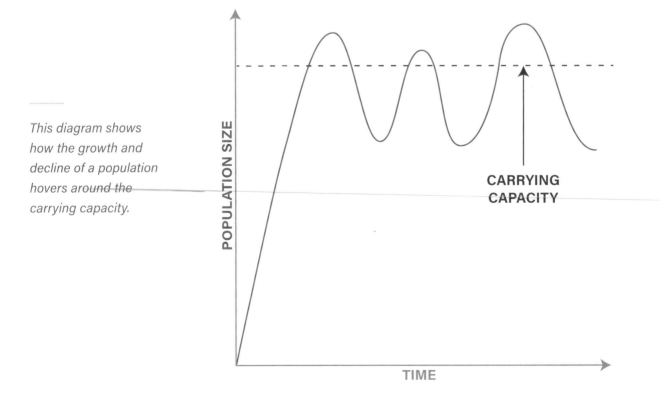

This diagram shows how the growth and decline of a population hovers around the carrying capacity.

OTHER FACTORS

Not all the factors that affect population size are connected with the population density. **Density-independent factors** affect population size, but do so *independently* from how many individuals are currently in the population. Examples of density-independent factors include catastrophes like floods and wildfires. Another example is habitat loss—sometimes humans fill in wetlands for development of shopping centers or other construction. This might affect a population of frogs, but the current number of frogs in the wetland doesn't make it more or less likely for it to be filled in. The development of a wetland would be independent of the current population size of frogs.

As you read in this chapter, some of the things that affect population growth or decline involve other organisms: competition with your own species or other species, being eaten by predators, or the population level of the prey species. In the next few chapters, we will explore the interactions between individuals and species, and include examples of how these interactions affect populations.

FOUNDATIONS REVIEW

- ✓ Population biology explores how the size of a population is affected by the survival, birth of offspring, and death and other processes of individuals within that population. To study populations, biologists might construct equations to describe the change in population size over time. The terms in a population equation would include how many births, deaths, and immigrations (new individuals coming in), and emigrations (individuals moving out) happen during a given period of time.

- ✓ Many factors can influence the size of a population. Some are called density-dependent factors, which means they depend on how densely populated a given area is. Examples would be the space and food available. Competition for resources (or lack thereof) can affect a population size, as can disease. For the most part, the size of a population will hover around the carrying capacity, which is a figure that estimates an average of how many individuals a particular habitat can "carry," or support.

- ✓ There are also density-independent factors that influence population size, such as catastrophic events like flood or wildfire, or habitat loss from human activity. These factors are independent of the density of a given population.

r-Selected versus K-Selected

In this chapter we learned about how population sizes fluctuate up and down in response to various pressures. Increases in food might lead to higher survival rates and more births, whereas disease or more predators could cause a population to crash. Understanding these dynamics helps us if we want to manage population sizes. For example, sometimes humans might cause population declines through unintentional harm, like through the effects of pollution or habitat loss. If an important or desirable species is affected, we could know how to help that population recover. Some organisms—those with faster growth or more offspring per year—will recover faster than others.

Within the diversity of life, we also see diversity in growth rates among species. Some have the "strategy" of having many offspring very quickly. Biologists call this r-selected, where the lowercase "r" refers to the growth rate. Even though we use the word strategy, animals or plants are not actually strategizing; they are not thinking through their decisions in the sense that we do. Species that are r-selected tend to have many offspring, little or no parental care (the parents do not stick around to protect or nurse them), smaller body size, and early maturity. Because there is no help for offspring, many do not survive to adulthood. Think of insects that may lay thousands of eggs, but only a few hundred make it to adulthood (this is a *quantity* over *quality* of offspring "strategy").

Other species have the strategy of having fewer offspring, but they provide more care, protection, and resources to their young. For example, elephants have one calf which is dependent on its mother's milk for two years! These organisms do best in stable environments, and are good at competing with other species. Biologists call these species **K-selected**, "K" standing for the carrying capacity. This indicates that these species are well prepared to

compete at carrying capacity. When resources are scarce, their care of their offspring helps them survive. But they also have a lower growth rate because they invest so much in each individual offspring but have so few offspring, so their population size would not be able to rebound quickly if their population crashed (in other words, the death of a few elephants is a much bigger deal than the death of a few insects).

In terms of fitness, both these ways of reproducing—r-selected and K-selected—have advantages and disadvantages. Certain species invest more in quantity (r-selected), whereas others invest in each individual offspring (K-selected), but you can't really do both—an insect cannot have thousands of young and nurture each and every one of them! We as humans are clearly closer to those organisms who are K-selected, since human parents have one child at a time (with the exception of twins, triplets, etc.) and are very involved in raising them. Perhaps it is for that reason that we feel a sense of admiration for and a connection to the animals that raise and protect their young, like a mother bear and her cub.

Battles take place everywhere in nature, including at the smallest levels, as these two stag beetles demonstrate!

CHAPTER 7

COMPETITION
Conflicts, Contests, and Battles in Nature

LIFE AND DEATH

As we have already discussed, living things need certain habitat requirements and resources in order to survive. In the early chapters, we focused on the abiotic (non-living) parts of the environment, like water and temperature. In the previous chapter, we started to explore other necessities—resources like food or nutrients which provide for individuals and help populations grow. But as populations increase, there are fewer resources available for each individual, and so **competition** is inevitable. Competition is a word you are probably familiar with—it describes how individuals might compete against one another for some desired goal. We humans compete in sports, and the prize might be a trophy. But in nature, organisms' competition is usually over life and death matters as they fight for food and other resources. In this chapter, we will explore how this competition resolves itself, affecting individuals, populations, and ecological communities.

Two red deer fight over a mate during rutting season (between the middle of October to early December).

ECOLOGICAL NICHES

Defining an ecological "niche" is a good way to start examining the interactions between individuals that affect populations and communities. An organism's **ecological niche** is the role it plays in the ecosystem. The ecological niche of an organism includes its environment and habitat requirements (space, temperature, nesting sites, etc.), as well as the interactions it has with other species. That is, what species does it compete with? What does it eat? What organisms eat it? An ecological niche basically includes all of the abiotic and biotic factors that affect an organism's life.

Consider for a moment what your ecological niche might be. This would be the home, neighborhood, and community you live in, as well as the family and friends you interact with. No one eats you (hopefully!), but you do probably eat chicken, beef, fruits and vegetables, and bread, among other things. And you do, in a sense, "compete" with your brothers and sisters for space in the house—you might need to use the bathroom at the same time, or share a room. All these many factors contribute to the ecological niche you occupy. You live in a certain home with certain people and you carry out certain activities and eat certain things.

Let's look at some examples from nature now, starting with some backyard songbirds.

The ecological niche of the Carolina wren (*Thryothorus ludovicianus*) includes the foods that it eats—mainly insects, spiders, and other small creatures—but the niche also includes the places where it finds this food. Carolina wrens primarily forage for food along the ground in leaf litter, along the base of trees, and in small cracks and crevices in the ground. If two species of birds in a backyard compete for these insects, we would say their ecological niches overlap (kind of like how you and your siblings' niches overlap as you try to raid

The Carolina wren (left) and the eastern towhee (right) occupy the same ecological niche and so often find themselves competing for resources.

your family pantry!). The amount of overlap is essentially determined by how much two species compete for resources, including food, nesting sites, refuge from predators, or whatever other resource an organism might need.

For example, eastern towhees (*Pipilo erythrophthalmus*) also forage along the ground for insects. Since they and the wren both eat the same foods found in the same spaces, any spider a towhee catches is one less spider for a wren to find and eat. On the other hand, the eastern towhee eats more seeds and plant matter and fewer insects when not raising young chicks at the nest (insects are a good energy source when parenting). This trait reduces their competition with wrens, or we could say it reduces the overlap in their ecological niches. We could also compare Carolina wrens with woodpeckers, like the downy woodpecker (*Dryobates pubescens*). Woodpeckers eat insects, but often capture them high in the trees, and from holes they excavate in the wood themselves (insects the Carolina wren couldn't access, since it doesn't have the tools to do so). So despite eating insects, woodpeckers may not overlap with wrens simply because they are pursuing different types of insects in different places.

TYPES OF COMPETITION

The kind of competition we just described is called **interspecific competition**, meaning two or more different species are competing (inter = between, specific = species). The competition we can observe at a birdfeeder with sunflower seeds is likely to involve interspecific competition, with many bird species enjoying sunflowers seeds, not to mention the squirrels or chipmunks that might join in. But any seed-eating bird would also be competing with members of its own species. The neighborhood population of black-capped chickadees (*Poecile atricapillus*), for example, might all visit the same feeder from time to time, competing for that resource. We use the term **intraspecific competition** to describe competition *within* a single species (intra = within, specific = species). To help you remember, think of *interfamily* competition as your family competing with a different family down the street in some kind of neighborhood game (between families), whereas *intrafamily* would be competition within your own family, perhaps over a board game.

In nature, both types of competition can happen at the same time and are not mutually exclusive, though it is true that one type of competition might be more important than the other, depending on the circumstances. If most of an organism's competition for resources comes from members of its own species, then intraspecific competition would have more of an effect on the individuals.

So far we described ecological niches and who competes with each other. But the *way* competition is resolved can also vary. Sometimes individuals compete directly, even fighting with each other, while other times the competition is more indirect. Let's use predators of the African savannah to look at some examples.

When individuals compete for a resource directly, we call this **interference competition.** A good way to remember this is that individuals directly *interfere* with others' ability to use the resource. This could be fighting for the best territory, or fighting for a food item or any other resource. A pride of lions (*Panthera leo*) and spotted hyenas (*Crocuta crocuta*) sometimes have physical confrontations at the carcass of a prey item (a zebra, gazelle, or other large herbivore). Neither group is willing to share the food, and so ultimately one or the other group is chased off. You might think lions would win that battle, but the outcome often depends on group size, meaning how many lions or hyenas arrived for the fight. Even a lion cannot overcome smaller animals if he is outnumbered ten to one (and so large carnivores can actually "count" how many

On the African savannah, a pride of lions and a pack of hyena battle. Though the lions are bigger, hyena are often able to outnumber them and come away victorious.

calls they hear to decide whether they should wander over to try to steal some food). Sometimes lions and hyenas also fight over space, chasing one another out of the best resting or hunting places.

Another predator of the African savannah, the leopard (*Panthera pardus*), is less directly aggressive with its competitors. Leopards, hyenas, and lions all hunt, kill, and feed on large animals. So all these predators living in the same area would compete for the local prey. Each prey captured and eaten is one less for other large predators. But even though they live in the same area, leopards tend to avoid places with lions and hyenas. For example, if hyenas were hanging out in the northwest part of a wildlife reserve, leopards might move to the southeast. As the hyenas move, so, too, will the leopard. So these predators compete for food but without direct confrontation (or at least, the leopards try to avoid such confrontation). We call this **resource competition**, or **scramble competition**. In this type of competition, the winner is the predator that catches the resource first.

SPECIALISTS VERSUS GENERALISTS

Competition can also be affected by how specific the needs of an organism are. Some species have specific requirements, like a narrow selection of foods to eat, or particular habitat requirements. In biology, we call these species **specialists**—they specialize on a specific set of resources. Koalas are often used as an example

Camera Traps and "Citizen Science"

Various experiments have helped scientists discover how large predators in Africa share or compete for space. Scientists often use camera "traps" to take photos of animals. These are hidden cameras covered or disguised somehow (so the animals will not notice them) that will automatically snap an image when movement passes before the lens. This can sometimes result in thousands of photographs. With so many images to sift through, the scientists ask citizen scientists to help! People of all ages log in to sort through photos and identify the animals found in each one.

of specialists because they only eat leaves from the eucalyptus tree. So koalas have the narrow diet of a specialist, unlike other species that eat a wide range of foods.

Much different than this would be another marsupial mammal, the wombat, which eats various grass species and other plants on the ground. Wombats are **generalists**, eating a broad, general assortment of foods. You may think it would be great to be a specialist, eating only your favorite food every day, but I'll join your parents in saying that the balanced diet of a generalist is a

Koalas are considered specialists since they almost exclusively live off a diet of eucalyptus leaves. Generalists, meanwhile, eat a wide range of things.

Any animal like this furry wombat that likes to munch on grass will almost always have food available to them.

healthier option for growing humans. For animals foraging in the wild, being a specialist often allows greater competitive ability, at least for the foods they specialize on (they are really good at getting that one food source). But generalists can avoid competition by opportunistically taking whatever food is most available. So, like other traits in nature, there is a potential trade-off between being a specialist and a generalist.

LEARNING TO LIVE WITH ONE ANOTHER . . . OR NOT!

So far, it should be clear that competing for resources is an important part of an organism's success, or its fitness. So much so that two species with extremely similar ecological niches may not be able to coexist in nature. We would expect the better competitor to push the other out of the area, or even to drive them to extinction. This process is called **competitive exclusion**, when one competitor excludes the other from the local area. Because competition is so important to fitness, organisms can also adapt to competition. If an organism can shift its diet, for example, to avoid competing with a better competitor, it would likely increase its food access (and thereby fitness) and avoid competitive exclusion.

Avoiding competition by reducing overlap in niches is called **niche partitioning**. This is like sharing the available resources, with each species focusing on a particular portion. You could use niche partitioning to share jellybeans with siblings or friends, everyone choosing a favorite color to eat. A classic example in ecology is the species of *Anolis* lizards in the Caribbean. Different species specialize in using different parts of their habitat to live and forage—along the ground, on shrubs, on tree trunks, or in treetops. On the islands this "space sharing" pattern is repeated, but often with different species filling an individual niche. For example, each island has a ground-dwelling species,

Species Fun Fact: Kangaroos can hop up to 25 feet in a single bound!

but different species can occupy the ground-dwelling niche. If a new species is introduced to the island, it would have the most success (highest fitness) if it fills an open niche (maybe there isn't yet a twig-dwelling species on this island), and so eventually all niches are occupied. The pattern of foraging in different areas reduces competition and allows coexistence of multiple species of lizards.

In our next chapter we will continue to discuss the conflict that we find in nature and how species will exploit one another ("use" one another), primarily by hunting and eating one another!

FOUNDATIONS REVIEW

✓ With a limited supply of space, food, and other resources, competition in nature is inevitable. Organisms must compete with one another to survive and reproduce.

✓ Each organism occupies an ecological niche, which refers to the role it plays in the ecosystem. The niche includes its environment and habitat requirements (space, temperature, nesting sites, etc.), as well as the interactions it has with other species. That is, what species does it compete with? What does it eat? What organisms eat it? An ecological niche basically includes all of the abiotic and biotic factors that affect an organism's life.

✓ There are various types of competition in nature. Some examples include interspecific competition, where different species compete for resources with one another, and intraspecific competition, where individuals of the same species compete against each other. Sometimes organisms can find ways to coexist by changing their diet or where and when they hunt—this is called niche partitioning. But when two organisms' niches overlap too much, the better competitor will push the other out (whichever is best at utilizing the resource or excluding the other will survive and claim the niche). This process is called competitive exclusion, where one competitor *excludes* the other from the local area.

A Priest and His Love of Ants

In this and the next few chapters, we will examine relationships between organisms, some that benefit only one of the organisms involved, some where both species benefit, and some where neither benefit. Competition, from our last chapter, is an example where no organism benefits. Yes, one organism might win the competition, but it would still be better off if it didn't need to compete at all.

There is another category of interaction we won't cover in great detail (other than briefly in chapter 9) called **commensalism**, where one organism benefits and the other is more or less unaffected. An example of this would be that many different species of beetles and other insects live in and among ant colonies. They benefit by gaining a home, protection, nutrition, or other things, but do not really bother the ants. This type of commensalism is called an **inquiline**—a species that lives in the home of another species without harming it. More specifically, an ant inquiline is called a **myrmecophile** (translates as "lover of ants"). Much of what we first knew about many myrmecophiles came from the research of a Jesuit priest named Father Erich Wasmann.

Father Wasmann was born in Austria in 1859. After becoming a priest, he became fascinated by ants and their interactions with other species. He studied how various species of ants interact and compete with each other, as well as the many insects that associate with ant colonies. He noted that many myrmecophiles evolved (developed over time) modified shapes, colors, and smells to imitate the specific types of ants they live with, as if they were trying to "blend in" as a member of the colony. We can picture selection acting on variation in these beetles, with those that "blend in" the best (smell/shape/color) being least likely to be kicked out by the ants. Over time, the populations became much more similar to their ant hosts and different from related beetle species. This type of imitation, or mimicry, to blend in with the host is still referred to as "Wasmannian mimicry." During his lifetime, Father Wasmann was recognized as a brilliant scientist and a leader in the dialogue between faith and science.

A fly caught in a spider's web is just one of many ways organisms are exploited by others.

CHAPTER 8

EXPLOITING OTHER SPECIES
Predation, Herbivory, and Parasitism

A WORLD OF EXPLOITATION

Using other organisms as a source of food or energy is called **exploitation**. Specifically, this term means that one organism benefits at the expense of another—it *exploits* (uses) the other for its own gain. Exploitation comes in various forms in nature, but in each interaction the fitness of one of the species is reduced. In this book, we have already used terms for some forms of exploitation, like predator, prey, or herbivore, but it may be useful to give these familiar words formal definitions.

A **predator** is an organism that kills and eats other organisms—typically animals that eat other animals.

Prey is a term to describe the organism a predator kills and eats.

An **herbivore** eats parts of plants, such as the leaves or stems. This harms the plant but often does not kill it. (Herbivory is the eating of living plant tissue.)

Parasites live on or inside a host organism, feeding on the body of the host, but typically without killing the host. But what is a host?

A **host** is the organism that harbors the parasite—essentially it is like the home for the parasite.

A **parasitoid** is almost like a combination of a predator and a parasite—it's an organism (often an insect) whose **larvae** (insect offspring) kill and eat the host.

A **pathogen** is a disease-causing organism (like viruses and bacteria).

There may be overlap between these definitions. For example, can we call herbivores predators if they eat the whole plant, or kill the plants they eat? Or if a parasite kills its host, perhaps it could also be considered a predator? But putting these tricky situations aside, one point of emphasis here is that the targets of exploitation (prey, hosts) are essentially under attack by other organisms. The prey or host is negatively affected in very significant ways. Unsurprisingly, being eaten by a predator is a terrible way to succeed in nature—it reduces your future fitness to *zero*. For this reason, we see many adaptations in nature to help organisms avoid these negative interactions.

Exploitation of other organisms can be as ordinary as a giraffe eating the leaves of a tree to as unusual as a wasp injecting her eggs into a hornworm so that when her offspring hatch they will have a nice meal in the caterpillar.

PREY TACTICS

Some of the more fascinating traits we find in nature function to deter predators, herbivores, and other exploitation. For animals, running away is one of the most obvious defense mechanisms, so long as they are faster than the predator! Better than running away, perhaps, is avoiding detection in the first place. **Camouflage**, or cryptic coloration, allows organisms to blend in with their environment. One great example of camouflage is the mossy leaf-tailed gecko (*Uroplatus sikorae*) which can look just like moss growing along a branch. Another form of camouflage is **disruptive coloration** which uses spots and stripes to break up the outline of an animal. For example, you may know zebras

The cryptic coloration of the mossy leaf-tailed gecko is almost too good to believe. Can you see him in the top image?

have stripes and giraffes have spots. But did you know this may help hide them? These patterns break up the outline of these large herbivores so they blend in with swaying grass, shadowy trees, and other vegetation. On the other hand, many predators rely heavily on smell, so this effect may be a bias of our own human reliance on vision! Research shows that zebra stripes may not be effective against predators, but instead may repel biting flies as they try to land—the light/dark variation seems to disrupt their landing process.

In addition to camouflage or disruptive coloration, organisms could also use mimesis to look like something uninteresting or non-threatening, like a leaf or twig, or mimicry to look like something dangerous—like caterpillars that have eye-spots to resemble a snake!

For organisms that can't run or hide, still other unique traits may help them avoid being eaten or attacked. For example, the pointy bits

SKUNK

(or "spikes") found on cactuses, porcupines, hedgehogs, pufferfish, and many other organisms make them less appealing. No one really wants a mouthful of spines! Physical traits like cactus spines are sometimes referred to as **structural defenses** (a term more often used in plant biology). Other organisms have **chemical defenses**—compounds in their tissues that are distasteful or even toxic. Various species of milkweed are one example of this, with bitter compounds that are poisonous (cardenolides, which interfere with the function of cells—cattle can even die from eating milkweed).

Animals and plants can have structural defenses in the form of spikes that ward off predators. Clockwise from upper left: blowfish (or puffer fish), small cactus plant, hedgehog, and the porcupine.

PREDATOR TACTICS

We see various traits that help plants and animals avoid being chewed on or eaten. But what can predators do in response? We can find ourselves feeling sorry for prey victims who die, but if predators cannot find food, they will also die. They need to eat to obtain energy to survive and reproduce, so predators often have adaptations of their own. For example, predators can also use the various forms of camouflage we described for prey. Prey can hide from predators but sometimes predators can also hide from prey. Some mantises and spiders camouflage themselves as flowers to catch insect prey. Leopards and jaguars have spots which act as disruptive camouflage. Other predator traits can be a direct response to prey defenses. Monarch butterflies lay eggs on milkweed, the plant mentioned a moment ago, and the caterpillars are able to sequester (sort of "save up") the toxic chemicals inside their own bodies. This means they are specialized to eat the poisonous plant and, instead of being harmed, become poisonous themselves!

COEVOLUTION

Predators and prey are locked into a sort of contest with one another, both species benefiting when either of them fails. This can lead to specialized deterrents in prey, and specialized responses by predators—like the milkweed's toxins and the monarch's ability to eat the plant without harm. This can lead to changes in the populations of each species, which we call **coevolution** (co = together, predator and prey populations change, evolve, in response to each other). One great example is the rough-skinned newt (*Taricha granulosa*) and the garter snake (*Thamnophis sirtalis*).

The rough-skinned newt and the garter snake have engaged in a long battle of counter-adaptations with each other.

The newt produces an extremely deadly neurotoxin (called tetrodotoxin), which paralyzes and kills predators. Garter snakes have resistance to the toxin, and so are able to eat the newts. But, more interestingly, in the parts of California where the newt populations have the highest concentrations of toxin, the garter snake population shows the highest levels of resistance! Imagine a newt population that varies in toxin, and a garter snake population that varies in resistance. In places where newts carry more toxin, those garter snakes with low resistance might be killed by the toxin. The next generations of garter snakes would have higher resistance, on average, simply because those with higher resistance would be the ones who survived and reproduced. In places with lower toxins, on average, high resistance wouldn't be needed for garter snake survival. Over time, this would lead to the matching pattern we see between levels of toxin and resistance in the two populations. These adaptations in contesting species, coevolution, are sometimes called **counter-adaptations**.

GARTER SNAKE

Counter-Adaptations

Let's look at a few more neat examples of counter-adaptations, where one organism finds a way to counter another organism's advantage as they engage in a predator-prey relationship. Bats use what is called echolocation to hunt insects. This is a way of locating prey (food) by using sound waves since they often hunt at night and have poor vision. The bats emit a sound wave which will reverberate back an echo. If that echo is disrupted or changed by the presence of something in the path of the wave, the bat knows something is there, in this case an insect. But, in response, some moth species have evolved the ability to make clicking sounds to confuse echolocating bats. The added noise makes it harder for bats to locate the moth.

Another fascinating example is how the crickets in Hawaii lost the ability to sing, or chirp (mentioned in chapter 5). Parasitoid flies lay eggs on male crickets, and the larvae eat the crickets from the inside out. Parasitoids are able to find the males when they sing to attract females. On some islands of Hawaii, crickets have stopped singing just in the last few decades. This helps them avoid being detected by parasitoids.

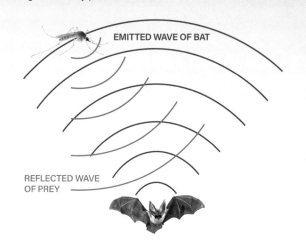

Remember:
Parasites live on or inside a host organism, feeding on the body of the host, but typically without killing the host.

PREDATOR AND PREY POPULATION EFFECTS

The interactions of individual predators and prey, or individual herbivores and plants, can also have effects on population size. In the earlier chapter on populations, we read about the factors that increase or decrease populations. When prey populations are high, there is increased competition among prey for food, which reduces the prey population. But at the same time, predators are eating the abundant prey. So their population might *increase* even as the prey starts to *decrease*, because predators are obtaining lots of food to fuel survival and reproduction. But you can probably guess that this pattern cannot continue indefinitely—with predators increasing and prey decreasing, competition among predators will increase, and some predators may end up starving. The predator population then crashes. In turn, this releases prey from predation pressure, more food is available, and so the prey population increases again.

It may seem that we are back where we started, and that's exactly right. Predator and prey populations often go through cycles, with predator populations following slightly behind prey. The classic example of the predator/prey population cycle is the lynx and snowshoe hare populations in North America. Scientists studied these populations extensively to help work out the ecological relationships between the two species. Interestingly, research shows that the plants the hares eat have an indirect effect on lynx populations. Lynx populations are highest shortly after hare populations hit their maximum. The prey maximum is largely determined by food availability and quality. This starts to show us how energy moves through a community of organisms—plants generate energy from the sun, hares eat plants, and lynx eat hares. A pulse in growth in the plants would eventually make its way up the food chain and increase the lynx population. It is fascinating to see how nature is connected in one big web. We'll learn more about communities and ecosystems in future chapters.

The lynx and snowshoe hare are connected through the food chain. If something happens to the population of one, it will affect the population of the other.

FOUNDATIONS REVIEW

✓ Using other organisms as a source of food or energy is called exploitation. The most obvious instance of this is predators hunting prey to "use" them for food. But other examples include herbivores eating parts of plants without actually killing them, or parasitic organisms living off a host organism.

✓ Prey animals have adaptations that allow them to avoid predators, such as being camouflaged or having disruptive coloration which uses spots and stripes to break up their outline, making them harder to see. But predators likewise use similar adaptations to hide from prey when they hunt. Plants, and some animals too, will use structural and chemical defenses to avoid being eaten. An example of this is a cactus's pointy bits (spikes).

✓ Interactions between predator and prey can have influences on the population sizes of both and can also cause counter-adaptations (or coevolution) in each, where traits are developed to counter the other's advantage.

Saint Augustine on the Nature of Animals

There is plenty of conflict in nature and even some gruesome outcomes for those involved: parasitoids that eat hosts from the inside out, parasites that suck blood and spread disease (like mosquitoes and fleas), or seemingly violent acts of predation. But these things are not evil as it is when human beings engage in violent behavior, and, in fact, in many ways nature's conflicts can be fascinating. Each animal, plant, fungi, or other organism is only acting according to its own nature, pursuing its own role in the larger ecosystem.

Saint Augustine teaches us that creation is good, even these less gentle parts of nature:

> This cause, however, of a good creation, namely, the goodness of God ... has not been recognized by some heretics because there are, forsooth, many things, such as fire, frost, wild beasts, and so forth, which do not suit but injure this thin-blooded and frail mortality of our flesh, which is at present under just punishment. They do not consider how admirable these things are in their own places, how excellent in their own natures, how beautifully adjusted to the rest of creation, and how much grace they contribute to the universe by their own contributions as to a commonwealth; and how serviceable they are even to ourselves, if we use them with a knowledge of their fit adaptations. (*City of God*, Book XI, Chapter 22)

> Attend, beloved. Who has arranged the limbs of a flea and a gnat, that they should have their proper order, life, motion? Consider one little creature, even the very smallest, whatever you will. If you consider the order of its limbs, and the animation of life whereby it moves; how does it shun death, love life, seek pleasures, avoid pain, exert various senses, vigorously use movements suitable to itself! Who gave its sting to the gnat, for it to suck blood with? How narrow is the pipe whereby it sucks! Who arranged all this? Who made all this? You are amazed at the smallest things; praise Him that is great. (Exposition on Psalm 148, 8)

A red-billed oxpecker takes a rest from flight on a willing partner in the impala, who has pesky insects picked off him by the oxpecker.

CHAPTER 9

TEAMWORK IN NATURE
Social Groups, Cooperation, and Mutualisms

Ecosystems Fun Fact: The flourishing of a healthy ecosystem depends in large part on the existence of mutualisms between organisms.

WORKING TOGETHER

The last two chapters showcased conflict in nature by focusing on competition and exploitation. Both of these factors are important ecological concepts. We saw that these interactions form an important part of the ecology of organisms. Not only do organisms need to deal with abiotic factors, like temperature and water availability, but they also need to respond to biotic interactions—predator/prey, parasitoid/host, and competitors for food or other resources.

But not all interactions between organisms are negative ones. Sometimes members of the same species, or members of different species, help each other out. When the individuals involved in an interaction both benefit we call this **cooperation**. A specific form of cooperation is a **mutualism**, which is when two organisms engage in a mutually beneficial relationship and do something helpful for the other. One form a mutualism can take is that of **symbiosis**, when two organisms live together in close association. In this chapter, we will explore these more "friendly" interactions in nature. Even though they may seem less dramatic than a predator stalking its prey, cooperation can also form an important part of an organism's ability to thrive.

COOPERATION

A common example of cooperation is when social groups work together toward some goal, whether a flock of birds, a pack of wolves, or some other group. Cooperative hunting is used by predators of various species. Perhaps the most commonly known are wild dog species, such as gray wolves (*Canis lupus*) and African hunting dogs (*Lycaon pictus*). These two packs coordinate to chase prey to exhaustion, using their numbers to keep the chase going. Many other animals also cooperate to hunt. One fascinating example is the brown-necked raven (*Corvus ruficollis*), which has been observed working in groups to catch lizards. First, two ravens swoop towards the burrow of the lizard, cutting off its escape. Only then do a few others fly down to catch the lizard. The whole enterprise seems planned out, and each individual seems to know its job. The burrow-blocking ravens only join their flock mates to share in the meal once the hunt is completed.

Packs of wolves (left) and wild dogs hunting together is a prime example of cooperation in nature. Each of these species has sophisticated tactics as they work together to stalk and capture their prey.

Snow geese will often send out alarm calls to warn others in the flock of potential danger.

Clearly, operating in a group can help predators, but it can benefit prey as well. There is safety in numbers. For example, each bird in a flock of one hundred individuals has less chance of being caught, both because there are so many other birds and because those others help keep an eye out for danger. Small songbirds also often work together to "mob" predators—a behavior where multiple small individuals "team up" to harass and chase away a dangerous predator, such as a hawk. One bird on its own might be vulnerable to a predator, but many species use **alarm calls** to gather others in the area to help. Mobbing behavior can include members of the same species but can also involve a mix of species because birds respond to other species' alarm calls. Each bird is taking a small risk, but all individuals benefit when the predator leaves the area.

Fish have been shown to cooperate in fascinating ways, including by investigating dangerous looking shapes together. In experiments, scientists will place a large predator-like shape in an aquarium, something that casts a looming shadow. The small "test fish" in the school inch forward a small distance at a time to check out the possible danger, and they will keep moving up as long as others in the school match their behavior. This means that fish only work together to check out danger if others are willing to help out!

While flocks of birds and schools of fish often work together, this does not mean they are permanent groups—individuals constantly leave and rejoin over time. So we wouldn't call this a mutualism, which is cooperation that requires continued close association between the partners.

Commensalism

Symbioses, cases of two organisms living together in close association, come in multiple forms. In a mutualism, both species benefit. We already examined parasitism, which is another form of symbiosis, where one species benefits and the other is harmed. **Commensalism** is a third type of symbiosis where one species benefits and the other is more or less unaffected. An example is the remora, a type of fish that attaches itself to sharks to get a ride, and feeds mostly on nutrient-rich shark waste. This helps the remora, but the shark neither benefits nor is harmed.

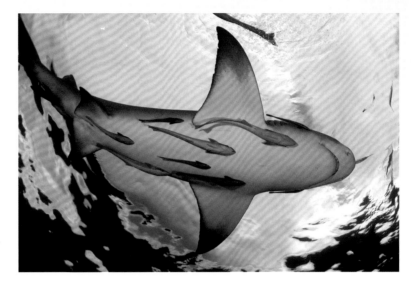

MUTUALISM

There are many examples of mutualisms in nature. The most common example is pollination—plants benefit from insects, birds, and other animals that carry their pollen from plant to plant. This fertilizes the plant, allowing seeds to develop, while the pollinator gets a sugary nectar reward, or sometimes eats some of the pollen. Other great examples include coral (consisting of a coral polyp and algae partner called zooxanthellae), lichen (a fungus plus multiple algae species as partners), or mycorrhizae (a mutualism between a fungi and plants that occurs at the roots, which is immensely important to many plants around the world).

Sometimes the partners in a mutualism depend on each other so much that they need each other to survive. Biologists call this an **obligate mutualism** (each species is "obligated," or required, to interact with the other to survive). Scientists can test if a relationship is an obligate mutualism by attempting to grow the partners separately—for example, trying to grow the algae and fungus

A coral polyp and its algae partner zooxanthellae (left) engage in a mutualism, but the algae and fungus from lichens (right) take it one step further, acting in an obligate mutualism, meaning one literally cannot survive without the other.

from lichens apart from each other (spoiler alert, in the lab, they fail to thrive independently). Alternatively, many mutualists are not entirely dependent on each other. Most examples of pollination are not obligate mutualisms. For example, a flower might be pollinated by a honey bee, bumble bee, or some other insect. The bees could pollinate any number of flowers. They can help each other but could also switch partners and survive without the other. This is called a **facultative mutualism**, or an "opportunistic" one.

COMPLICATED COOPERATION

So far, we have mentioned the various ways in which cooperation and working together in groups can benefit individuals. But often biological relationships are not that simple. Previous chapters highlighted conflict in nature, and this can still exist here within cooperative groups. For example, flocks of birds may benefit by reducing predation when they coordinate their movement and stick together. But being in a group could also lead to increased competition for food, or aggression from flock mates. Just like you may squabble with your siblings from time to time, so do animals. Other examples of cooperation and mutualism can also be complicated.

There are small species of fig-wasps that pollinate various fig species. The fig plant needs the wasps, as they are the only pollinators—no other flies, bees, or wasps are able to successfully pollinate the plants. But what does the fig wasp get out of this mutualism? The wasps lay eggs inside the figs, and the young

Fig plants and the fig-wasps have a complicated relationship that benefits both species.

eat parts of the fruits. On the whole, the fig plant still benefits (so we can still call this a mutualism), but the wasp isn't just a pollinator since the larvae (the wasp offspring) also eat part of the plant. Other mutualisms also involve a little give-and-take. There are various species of yucca plants, and each has its own specialized yucca moth species to pollinate it (each yucca-to-moth pairing is an obligate mutualism). The female moth transfers pollen to the yucca flower and then lays her eggs in the flower. Developing moth caterpillars then eat some of the yucca seeds, so the yucca moth is a mutualist and seed predator.

Here's another complicated web of interaction. Aphids are a type of insect that feed on sap from certain plants. Ants often defend and care for aphids because the aphids feed the ants a sweet honeydew liquid. The aphids get protection, the ants get a good meal. So this mutualism between ant and aphid can increase the population of aphids because the ants protect them from predators.

It would be reasonable to assume that more aphids around can be bad for the plant, since more of them are feeding on it. But research also shows that ants can help keep the plants clean of mold or fungus, which could be more harmful to plants than aphids. So maybe having ants around isn't so bad for the plant after all. To complicate the situation further, we could add in ladybugs and other predators that try to eat the aphids, or another ant species that eats the aphids and attacks other ants. All of these species would affect each other, even if they don't directly interact, and we're only looking at a small community on a single plant—imagine what interactions we would find in an entire forest, grassland, or desert! Individual relationships, like any predator/prey, or mutualism, are not isolated from others. In future chapters, we will start "scaling up" to look at larger levels of the ecosystem.

FOUNDATIONS REVIEW

✓ When the individuals involved in an interaction both benefit we call this cooperation. A specific form of cooperation is a mutualism, which is when two organisms engage in a mutually beneficial relationship and do something helpful for the other. One form a mutualism can take is that of symbiosis, when two organisms live together in close association.

✓ Predators like packs of gray wolves engage in cooperative hunting, while flocks and herds of prey also find safety in numbers by protecting each other through sheer numbers or by giving off alarm calls for danger. Sometimes the partners in a mutualism depend on each other so much that they need each other to survive, which biologists call an obligate mutualism, while other relationships that are not as imperative for survival are called a facultative mutualism.

✓ Interactions among individuals that seem helpful at first may also present complications. For example, while it may offer protection for prey to be in a flock or herd, it also means there is more competition among them for food.

Understanding Anthropomorphism

In this chapter, we learned how organisms work together in mutualisms, acting in ways that benefit all the partners in the mutualism. When organisms work together, it can appear to us as if they are being kind to one another, or that they want to help each other out. A particular form of help, usually within the same species, is called **altruism**. This is defined as helping another individual at a cost with no personal and immediate benefit for yourself.

An example of this might be the cooperation of vampire bats. They share their blood meal with other hungry bats in the same roost, even unrelated individuals. Over time, sharing food builds up "friendships" between bats, making them more likely to share with each other in the future. Another example is a type of fish from Antarctica which "adopts" abandoned nests. Males care for nests of young, and, when a nest is abandoned or another male is killed, an unrelated male may care for the nest.

In examples of altruism and helping in nature, it's easy for us to assign human thoughts or feelings to the helping behavior. This is **anthropomorphism**, which is giving human-like characteristics to non-human animals. We often do this because we experience thoughts and emotions, and so end up using our own experiences to try to understand other organisms. For the fish adopting nests above, this is perhaps just an extension of the fish's instinct to care for its own nest, as animals do not have consciences, so no voice urging them to act out of love.

Only human beings have a spiritual soul and an intellect. We can choose to care for others because it is morally right. As Christians, we are called by Christ to love our neighbors as ourselves. Animals do not have a soul or intellect, and so are not called to this same teaching. Even so, we can still bear a love for all kinds of animals and appreciate their natures, and when we do so, we give glory to God!

A pack of spotted hyenas attack a flock of white backed vultures as they prowl for a meal. But the food chain goes both ways, as these predators could tomorrow be the prey of a pride of lions!

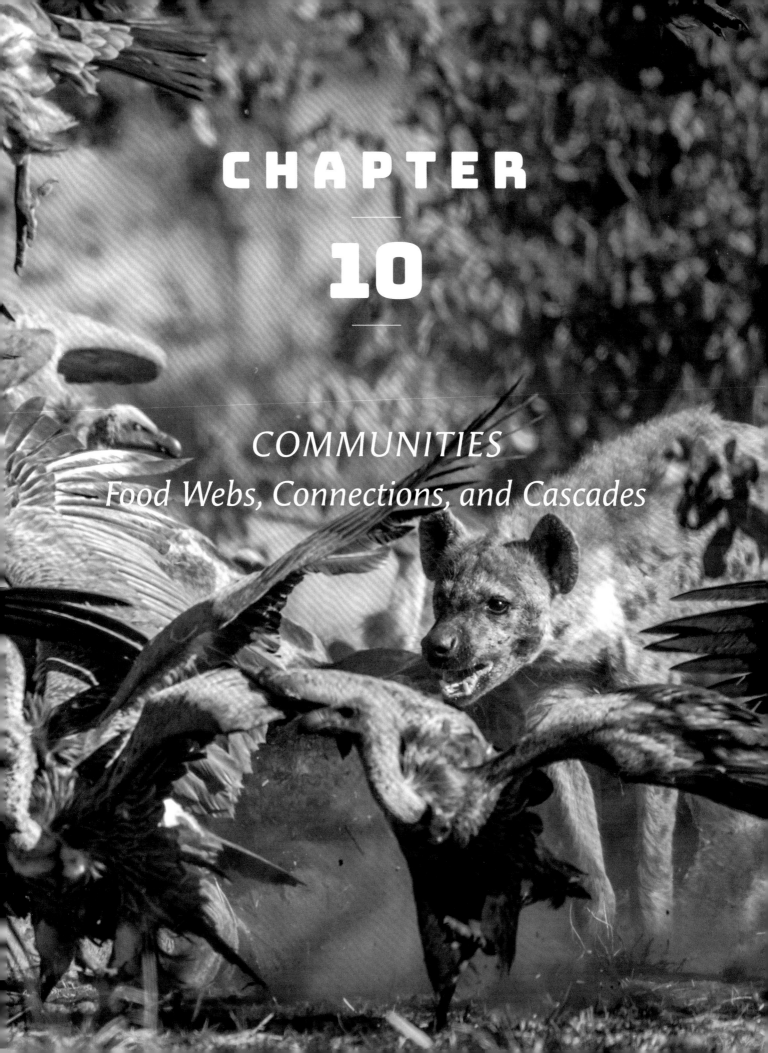

CHAPTER 10

COMMUNITIES
Food Webs, Connections, and Cascades

INTERACTIONS IN AN ECOLOGICAL COMMUNITY

All of the interactions between species, called interspecific interactions as we discussed in the last few chapters, form component parts of an ecological community (also called a biological community, or simply, community). In ecology, an **ecological community** is the set of biological organisms that occupy and interact in a particular area. In this chapter, we will build on our understanding of the interspecific interactions (predation, parasitism, competition, etc.) by studying them in the context of the larger community.

Many of the interactions we've explored thus far involve foraging and food, in other words, who eats whom, and which species compete for food. Instead of studying pairs of species at a time, we can start to build a **food chain** by drawing connections based on which species eat other species. For example, a frog eats a fly, a raccoon eats the frog, and a bobcat eats the raccoon. We could add in other species that eat frogs, such as herons or snapping turtles. And the raccoon also eats fruits, fungi, crayfish, and many other things. Soon we have an interconnected **food web**, showing us many of the interactions within a community.

As this diagram shows, "food web" is a more accurate term than "food chain", as there is no single chain that works its way up and down nature, but rather a complicated web of connections. Still, food chain is the more common term.

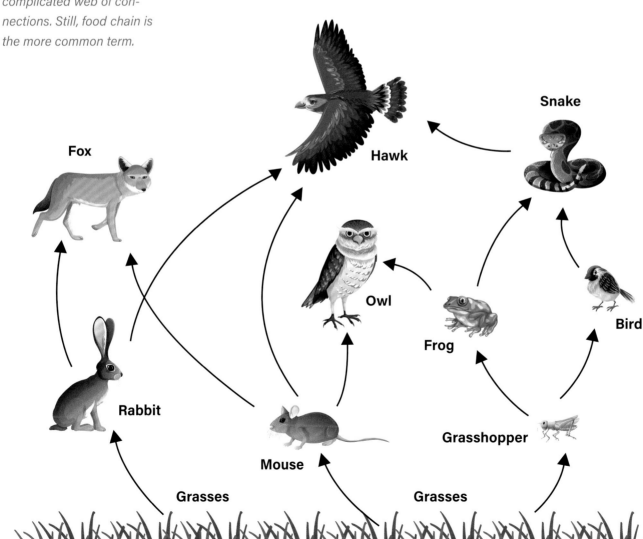

FOOD CHAIN

You can see how this would quickly get complicated, so scientists like to categorize things and break them up into smaller parts, which helps us better understand the whole. Let's take a look at several ways scientists divide and define communities and food chains or food webs.

TROPHIC LEVELS AND GUILDS

One way to represent species is by **trophic levels**, which essentially show where a species sits on a food chain. Put another way, it is how many steps they are from the sun and photosynthesis, since that is the source of all energy. Plants or other organisms that conduct photosynthesis are **producers**. Organisms that eat other organisms are called **consumers**. We can show how far away the consumer is from the producer using the following terms:

- Primary: One step, the herbivores
- Secondary: Two steps, carnivores
- Tertiary: Three steps, higher carnivores
- Quaternary: Four steps, highest carnivores

So a rabbit eating grass is a primary consumer, meaning it eats the producer directly. A fox that eats the rabbit is a secondary consumer, and a coyote that eats the fox is a tertiary consumer (though a coyote may also eat the rabbit). If a grizzly bear ate the coyote, it would be the quaternary consumer. At the very top of each food chain sits a predator like this, or an **apex predator**, that has no predators which eat it (nothing would dare hunt the grizzly bear!).

QUATERNARY

TERTIARY

SECONDARY

PRIMARY

"Pertaining to Food"

In biology, the word trophic, or word endings of -troph, indicates something about food. The words come from the Greek word *trophikós*, which means "pertaining to food." Trophic levels show where an animal is in a food diagram. Other quick examples include autotrophs, which are organisms like plants that generate their own food (auto = automatically make; troph = food), while heterotrophs eat other organisms (hetero = another, so eats another to get food).

Similar to trophic level is a **guild**, which is a set of organisms that exploit the same or similar resources. Whereas trophic levels are any organisms found at a particular level in the food chain, there would be multiple guilds found at each level. Take primary consumers, for example. Mice that eat seeds, or rabbits that eat grass, are both primary consumers. But we would put them into two different guilds—we could call them seed predators and grazing animals (not all guilds have formal names, instead we can just use descriptors). Within a group of grazing animals, scientists might focus on different parts of the grasses. One guild might consist of animals that eat the blades of grass, while another consists of animals that eat down to the roots or rip up whole plants.

Organisms in the same guild will often (but not always) share similar ecological niches, and thus compete for resources. You can easily imagine how different rodents that all want to eat seeds would directly compete with one another to find and collect their food. But there could also be specialization on certain types of seed—maybe the larger rodents focus on the biggest ones, for example. This would reduce competition. Or, from our grass example, certain animals might focus on particular types of grass, which we could use to further subdivide the "grazing" guild. This focus on different types or species of grass would allow the grazers to reduce the overlap in their ecological niche and reduce competition between them—each species can benefit from this reduced competition and this can allow coexistence.

A herd of zebras graze on the African grass, while a giraffe enjoys a meal in the trees. The giraffe's long neck allows it to be free from competition with the zebras and other animals that might eat the grass down low.

TROPHIC CASCADES

Communities of organisms are linked together through connections between trophic levels and guilds. Using the images of chains and webs to describe communities of organisms is useful because it illustrates these connections. Effects from predation, parasitism, and competition not only affect individual species involved in those interactions but also other members of the community. For example, the loss of one species in the food chain will affect its predators (and anything that eats them), and its competitors (and anything that competes with or eats them). In other words, a single change in a community can ripple through the food web, like one domino falling to knock over the next. Scientists call these effects a **trophic cascade**. Cascade is a word used to describe something being passed on from one thing to the next, or it can also refer to a kind of waterfall—the water runs or passes through the river and over the cliff. This is a good image to use because the change in the community has an effect that "cascades" (runs) through the different trophic levels.

Trophic cascades generally show up in two ways: top-down or bottom-up. The original description of a trophic cascade was by an American ecologist named Aldo Leopold, who worked in the early 1900s. He observed a top-down cascade when he noticed that wolves removed from a mountainside community led to more grazing animals (since the wolves were not there to hunt

Aldo Leopold sits on the edge of a rockface in Mexico. His work in ecology in the early 1900s helped us understand the phenomenon of a trophic cascade.

them), which killed off grass species (since more animals were eating the grass). Here the wolf is at the "top" of the food chain, and removing it has an effect that cascades down to affect grasses. Wolves don't eat grass but, by consuming herbivores, they indirectly help grass species.

A bottom-up cascade is when something happens at the bottom of the food chain which affects the levels higher up. So if the grass in that same community died from drought, that would lead to fewer grazing animals (they would die without grass to eat), and then the wolves would also go hungry with fewer animals to hunt. The cascade is running up instead of down.

As a general example, fewer producers conducting photosynthesis inputs less energy into a food chain, which would ultimately lead to fewer consumers at various levels. Because energy transfers up the food chain, but does so inefficiently, there are always more producers than consumers, and so this bottom-up effect is always present. In fact, the food webs found in communities across the world tend to have similar ratios of each trophic level. For example, one study showed that prey species tend to outnumber predators at about a 2.5 to 1 ratio (2 to 3 prey species for each predator species present in a food web).

APPARENT COMPETITION

As we can see, species can have indirect effects on one another through food web connections and trophic cascades. **Apparent competition** is a related phenomenon, where species that do not directly compete still affect each other indirectly through their relationship with another species, such as in having a common predator. We could have lumped this into the chapter on competition, but in this

Ecosystems Fun Fact: An interesting effect from the inefficiency of energy being transferred up the food chain is that there are limits to how many links can be in a food chain. The average number of links (from producer to the top predator, counting each step in-between) is about five, and they rarely, if ever, get to ten.

Remember:
In biology, the word trophic, or word endings of -troph, indicates something about food.

type of interaction, there is no *direct* competition. A good example of an apparent competition comes to us from an experiment that studied the effects of a mustard plant that was accidentally introduced into native plant communities in the United States. It seemed that the mustard could outcompete native grass species, taking over the area. But researchers studying the system showed that the biggest effect wasn't competition between plants; instead, mustard provided a safer shelter for mice to hide from predators. In turn, this increased the population of mice, which prefer to eat grass seed. So it looked like the mustard plant was driving out the native grasses, but it was not doing this directly—only because the mice were surviving longer and thereby eating more grass seeds.

Studying these intricate systems in nature is fascinating. There are so many different connections between species in communities that sometimes it can be hard to disentangle the food web to study it. As we continue our ecology exploration in the next chapters, we will focus on the attributes of communities and how they change over time. This will include some examples of how we can show our care for God's creation and these wonderful communities and ecosystems!

FOUNDATIONS REVIEW

✓ An ecological community, also called a biological community or simply community, is the set of biological organisms that occupy and interact in a particular area. Many of these communities are organized by scientists through food chains and food webs (who eats whom).

✓ One way to represent and classify species is by trophic levels, which essentially show where a species sits on a food chain. Put another way, it is how many steps they are from the sun and photosynthesis, since that is the source of all energy. Plants or other organisms that conduct photosynthesis are producers. Organisms that eat other organisms are called consumers, which can be classified as primary consumers (one step away from the producer), secondary consumers (two steps away), tertiary consumers (three steps), or quaternary consumers (four steps).

✓ A single change in a community can ripple through the food web, like one domino falling to knock over the next. Scientists call these effects a trophic cascade. Cascade is a word used to describe something being passed on from one thing to the next, or it can also refer to a kind of waterfall—the water runs or passes through the river and over the cliff. This is a good image to use because the change in the community has an effect that "cascades" (runs) through the different trophic levels.

Therese von Bayern
A Catholic Scientist

What comes to mind when you think about the life of a princess? If you said dangerous expeditions to unknown territories, courageous determination and strength, diligent scientific studies of communities of organisms, and a strong Catholic faith, then you would be describing the life of Therese von Bayern! She braved malaria, broken bones, difficult climates, and much more to study God's creation.

Therese was born in 1850 into a royal family in Bavaria, a region now found in southeast Germany. Her father was the Prince Regent, or ruler of Bavaria, from 1886 to 1912. During her childhood, she always had a thirst for knowledge, which is perhaps what drove her expeditions later in her life. Amazingly, Therese learned twelve languages during her lifetime! Her family, especially her mother, also instilled in her a strong Catholic faith. Therese kept journals throughout her life which detail her faith in God. Her life is another great example of how studying science is not incompatible with a deep faith.

Ecology was not really defined as a discipline within biology during Therese's lifetime, but her interdisciplinary interests might define her as an ecologist. She was interested in many topics, including geology, geography, botany, and zoology. She once led a team of four people on an expedition through Brazil. They traveled hundreds of miles along the Amazon and the coast of Brazil, making scientific observations and collecting many organisms (both plants and animals, including many butterfly species). She carefully documented every specimen and wrote a book about the scientific observations from her expedition.

Due to her contributions to science, including documenting the biodiversity of Brazil, she was the first woman elected to the Bavarian Academy of Sciences, and was the first woman to receive an honorary doctorate from the University of Munich.

A spotted fire salamander sits atop a cep mushroom on the forest floor.

CHAPTER 11

BIOLOGICAL COMMUNITIES AND BIODIVERSITY

KEEPING COMMUNITIES STABLE

Our last chapter helped us understand how individual interactions "scale up" to build a biological community. In this next chapter, we will look at some of the questions ecologists ask about communities, like how they function together, and how they change over time. We will also see how some key species can play important roles in maintaining communities and stabilizing their own environments.

As we study biological communities, one thing we might be interested in is the stability of a community, or how well it persists or maintains itself over time. For example, we rely on various natural communities for **ecosystem services**, which is a way that nature helps humans. Bees pollinating crops, bacteria removing waste from the environment, or wetland plants preventing flooding are all good examples. If we want to ensure these benefits continue, we would want to know that the biological community that provides these services will persist (meaning survive and remain unchanged).

One of the factors that affects the stability of a community is its **biodiversity**—that is, the diversity of organisms found there. A well-established community might be stable and balanced with the right mix of species in various niches and guilds, but sometimes things happen to shake up this balance. These challenges could include things like climate change, drought, removal of a species, or any other changes to the environment or the species present. Scientists predict that diversity helps communities persist in the face of challenges because with diversity there would be more links in the community's food web. For example, maybe one flower species is lost during a drought, but if there are many types of flowers, the other species could help fill the gap. Diversity is like a backup plan for when changes occur.

The idea that many species would make for a more robust community might make intuitive sense to us—more species, more chances that the loss of a single species won't cause a catastrophe. But scientists can also test this by creating experimental communities with different numbers of species (like some with only one or two plant types, and others with many). These sorts of experiments tend to support the hypothesis that biodiversity helps communities be resistant to disturbances and change. They also suggest that the occurrence of more species in an environment leads to higher **productivity**, or more growth and life (more on productivity in a moment). If an environment has more plants and other producers, this leads to more consumers, which all equals a more biodiverse environment.

MEASURING BIODIVERSITY

So how do we measure species diversity? There are several metrics that scientists use. The simplest is **species richness**—this is just a count of how many species are present. You could go outside and carefully count the types of insects you find to measure the species richness of your backyard insect

Ecosystems Fun Fact: Did you know half the world's species live in tropical rainforests?

Mangroves grow deep within Everglades National Park in Florida. This is one of the most biodiverse ecosystems in the United States because of its warm and wet climate.

community. But richness does not account for how many *individuals* of each species are present, meaning which species have large populations or small ones. As you would expect, there are often species that are extremely abundant and some that are rare. For example, wetlands in the southeast United States have many mangrove trees. We would call these trees the **dominant species** of that ecosystem because they are abundant and important to the community.

There are various ways to account for the abundance or rarity of each species in measures of biodiversity—we call these a **diversity index**. One example is **species evenness**, which aims to show whether there is an even distribution of species, or if there is one (or a few) extremely dominant species. Let's look at a hypothetical example.

Ecosystem 1	Ecosystem 2
Contains 3 different tree species	Contains 3 different tree species
Species 1 = 90	Species 1 = 33
Species 2 = 5	Species 2 = 33
Species 3 = 5	Species 3 = 34

Which one of these would you say has higher (or better) species evenness? If you said Ecosystem 2, you're right! It has the closest to an even count as we could get in either ecosystem, while Ecosystem 1 has low evenness.

THE BENEFITS OF A COMPLEX ENVIRONMENT

Once we have a way to measure or quantify biodiversity, scientists can also ask questions about what factors increase or decrease diversity. One factor that leads to higher biodiversity is **complexity** in the environment, meaning a high variation in species and niches in a given environment. When we learned about competition, we saw that niche overlap can lead to species competing. A limited number of niches means that only a limited number of species "fit" in a particular environment.

But we often see quite wonderfully diverse natural places around the world. In a forest, perhaps more sun reaches the forest floor in various patches, and some soil is wetter because water runs downhill. These patches of forest with

Remember:
Biodiversity is a measure of how diverse a community of organisms is in a given habitat.

slight differences in sun, water, nutrients, and microbes in the soil allow a greater variety (greater complexity) of plants to grow. More plant types then leads to more animals. Sometimes we also call this **environmental heterogeneity**, which is just a way to say that places—like a forest or a prairie—don't look exactly the same everywhere (they are complex). Next time you take a nature walk, take a look to see if you can find patches of plants that are different from each other. You could also take a closer look to see if you find different types of insects along your walk.

MAINTAINING BIODIVERSITY

When we look at species abundance in a typical ecosystem, most species have an intermediate abundance, meaning average, or in the middle. But there will also be a few highly abundant species and a few rare species. As we noted a moment ago, highly abundant species are called dominant species. Dominant species are important to their community because they hold a lot of the **biomass** and energy available in that system (biomass = the added sum weight of living things in that area, which can be divided by species).

But does that mean rare species are unimportant? No! In fact, sometimes rare species can play extremely important roles in a balanced ecosystem. Species that have low abundance but have an important effect on their community

BIOLOGICAL COMMUNITIES AND BIODIVERSITY **97**

MOUNTAIN LION

are known as **keystone species**. Good examples of keystone species include top predators, like gray wolves and mountain lions. If these keystone species disappear from an ecosystem there can be dramatic effects.

How might this happen? Well, imagine that human beings move into an area and build homes. They will want dangerous animals removed to make it safe. If top predators are hunted and killed or captured and removed, and therefore are lost from that ecosystem, their prey will no longer be chased and eaten (those top-down cascade effects on the community are removed).

This means they can go wherever they want and that their population will increase. In several national parks, when certain predators were removed (perhaps to make it safe to camp or hike), elk and other large herbivores ate in open areas near streams without the threat of being killed or eaten themselves. The elk eat young tree saplings, which reduces beaver populations because the beavers compete for the same saplings. The corresponding loss of beaver dams, and the ponds they create, led to less biodiversity overall. Everything in this web of nature is connected. So finding ways to coexist with these large predators can lead to more balanced, beautiful national parks.

Speaking of beavers, they are another great example of a single species that has a large impact on its community. We call them **ecosystem engineers** because they alter their own habitat. This also has the effect of creating habitat for all sorts of plants and animals that utilize the ponds they make—whether aquatic critters living in the pond, or those that use the pond for food or drink. Other examples of "engineers" include prairie dogs which create tunnel systems that animals can use as shelter and also affect plant life; or woodpeckers which make nesting holes in trees, which are then later used by wood ducks and other species (who can't make their own nesting holes).

Nature is full of wondrous diversity, connected together in populations, communities, and ecosystems, all of which glorifies God. As we heard from Saint Augustine in the introduction to this book, we give

(Below top) When top predators are removed from an ecosystem, animals like this giant bull elk are allowed to roam more freely, which can lead to a domino effect of changes. Beavers (below left) and prairie dogs (below right) are examples of ecosystem engineers, building dams and tunnels that other animals benefit from.

Partnering with Beavers

A research project at Utah State University has some unique research partners—beavers! The goal is to help restore stream health in Western states. Some small streams have started drying up in summer months, but beaver dams built along the course of a stream can help keep water in it year-round (storing it up behind ponds). Reintroducing beavers, or having students mimic the building of beaver dam structures, can result in a reliable source of water for wildlife and livestock!

creation a voice to praise Him by studying it and marveling at its beauty. The next chapter will close our book by scaling up our attention to the level of ecosystems and challenging us to consider how we might also praise God by being good stewards of His works.

FOUNDATIONS REVIEW

✓ One of the factors that affects the stability of a community is its biodiversity—that is, the diversity of organisms found there. A well-established community might be stable and balanced with the right mix of species in various niches and guilds, but sometimes things happen to shake up this balance. Scientists predict that diversity helps communities persist in the face of challenges because it creates more links in the food web.

✓ Measuring biodiversity helps us study it and ultimately protect certain ecosystems. We can measure biodiversity by keeping a biodiversity index, that is, keeping track of how many species (and how many individuals within each species) live in a certain environment. Some species will be average, some will be abundant, and some will be rare, but even the rare ones can have important roles to play. Some of these rare species may be keystone species, meaning they have an important effect on the community.

✓ Complex environments, or ones that have environmental heterogeneity (that is, a wide variety of species) are not only beautiful, they are also better able to survive and maintain their biodiversity. Having an abundance of species means if one species dies or is removed, there may be others that can "pick up the slack" and step in to fill its role. But sometimes the loss of a certain species can cause dramatic cascade effects that expand through the entire ecosystem.

Invasive Species

Throughout this book, we have observed various ways in which nature is interconnected and also operates in a balance. For example, predators keep nuisance organisms in check (as bats do with mosquitoes), predator and prey populations go through cycles, energy and nutrients move through the food web, and more. But sometimes this system is thrown off-balance. The most dramatic example may be habitat destruction, like when human beings clear out a forest to build neighborhoods or shopping centers.

Another example of imbalance comes from invasive species. An **invasive species** is an organism that is not native to a particular region or habitat, but has been introduced there, primarily by humans (either intentionally or unintentionally). Why is it an issue if new species arrive in an area?

Well, in the chapter on predation, we saw that predators and prey can co-adapt with each other—but new predators might have a distinct advantage. For example, islands without rodents have birds that make their nests on the ground because there is no danger to their eggs or chicks. But if certain rodents find their way onto those islands, these ground-nesting birds will not be prepared to defend themselves or their nests from scurrying predators looking for a meal. Various albatross species are at risk due to the introduction of rats to their communities. Danger from mammals was never an issue, and so their instincts and behavior were mismatched to this new threat.

Other examples of invasive species include plants or insects that are released from their natural predators. If nothing eats them in their new habitat, the population grows unchecked and outcompetes native species that are still being eaten by their natural predators. Due to these various issues, invasive species can reduce biodiversity (by excluding native species) rather than adding to it. Because of the potential for damage to human infrastructure, valuable or harvested species, or entire ecosystems, lots of research is ongoing to determine how to control these pests.

One of the great challenges to our role as stewards of the Earth is to make a way of life for ourselves while still being mindful of the delicate balance of God's creation.

CHAPTER 12

ECOSYSTEMS
Finding Balance within Communities and Environments

THE INTERCONNECTEDNESS OF ECOSYSTEMS

In our last chapter, we shifted focus to the larger scales, or levels, of ecology. Now we will review our definition of ecosystems, see how they can change over time, and discover how we can protect or restore natural systems. By exploring ecosystems, we will also see how scientists assess ecosystems, or measure their function and stability. Finally, we will look at how ecosystems connect to one another in larger landscapes.

In the beginning of the book, we examined how abiotic factors like temperature and water availability affect individual organisms. Other abiotic factors that affect living things include acidity (the pH of a system), salinity (level of salts), the light energy available, or various attributes of the soil (nutrients, chemicals, or water content). When we consider all of these abiotic factors of a given place together along with the community of living things found in that area, we have an ecosystem.

We have already explored interactions on smaller scales throughout this book—interactions between an individual and its environment, or multiple individuals competing, or populations of species affecting the abundance of other species. All of these interactions scale up and come together to provide an interconnected ecosystem. The "system" part of the word ecosystem reminds us that it operates as a *unit*, and changes to one aspect of the ecosystem affect other parts.

For example, water availability may fluctuate over time as a water cycle distributes and moves water (through precipitation, evaporation, etc.). This could cause periods of drought or periods of flooding, which in turn affect the living things, shifting the makeup of communities. Other aspects of the abiotic environment also change over time. Nitrogen content in the soil could change, affecting nitrogen transitions to herbivores as they eat plants, then to carnivores as they eat herbivores. When living things die and decay, decomposers like bacteria and fungi help return these nutrients to the soil. These cycles (which are also described in the *Foundations* unit on planet Earth) help show how the ecosystem is interconnected.

Abiotic factors like rainfall can drastically affect ecosystems, with periods of drought and periods of flooding.

We can look at the energy flow or nutrient cycling of an ecosystem to understand its health and stability. Nutrient cycling (movement of carbon, nitrogen, or other nutrients through the ecosystem) and energy flow (conversion of light energy by plants, and then transfer of that energy up the food chain) are both properties of the system as a whole. Missing links in the food chain, for one example, could disrupt this flow. So by understanding these processes, we can get a deeper understanding of ecosystems or identify where issues might exist.

Overall, these cycles of nutrients and energy are a part of **ecosystem function**—or the set of processes that are necessary to maintain the living system. Included in ecosystem function are **ecosystem services**, or the goods and services we get from nature (food, medicines, flood management, erosion

Healthy food is one of the most important ecosystem services we receive from God's wonderful world of nature.

control, water purification, and so much more). Maintaining healthy, functioning ecosystems is not only part of our stewardship of creation but also helps us care for ourselves and others by maintaining these ecosystem services.

Productivity is another part of ecosystem function, which is the generation of biomass—essentially, how efficiently producers turn light energy into growth. This measurement is important because it affects how much energy is then available to the rest of the food chain, including us. Perhaps studying ecosystem productivity can also help us understand how to increase agricultural productivity and increase the biomass (food!) we can grow.

CHANGES IN ECOSYSTEMS

As we have seen, healthy, functioning ecosystems can be important for various reasons. But if parts of the system change over time, do ecosystems themselves change? The answer to that is generally yes—but the answer may change depending on how we look at it. First, the timescale could affect our answer. For example, pine forests far north in Canada or Alaska and undisturbed by humans may seem unchanging for decades. These would be beautiful places to visit, and you might not see dramatic changes if you took a vacation there each year. But many decades or hundreds of years ago, some of these places were covered by glaciers of ice. So over long periods of time, there has been a dramatic shift in the environment and also in the organisms found there. This process of ecosystem change is called **ecological succession**. Succession is a word that refers to one thing following, or changing from, another thing that came before it. In places where glaciers receded, the changes go through similar, predictable changes—first shrubs are dominant species (within decades), then trees such as alders, and finally the pine forests (over hundreds of years). The animal community changes over time too, as different animals can live in each plant community.

These two images of a pine forest in Estonia show the phenomenon of ecological succession. The image on the left is one year after a forest fire, while the one on the right was taken two years after the fire, where we can see more vegetation starting to grow back.

Once the pine forest is established, the community may seem stable to us as we visit. But though the ecosystem and its functioning may be stable at a large scale (the ecosystem level), the community assembly may change over time, or populations of individual species may fluctuate. For example, certain songbird species may increase as they outcompete others for seeds, leading other species to decrease. One seedeater may replace another, or a grass species may be replaced by others. Some long-term ecological studies demonstrate these effects, showing that an ecosystem can remain stable even if changes occur within the community (in other words, the "big picture" remains relatively stable even if several "little pictures" change within it). In a way, the ecosystem itself is an active system, responding to changes over time.

OUR OWN ROLE IN PROTECTING ECOSYSTEMS

Succession occurs over long periods of time, but some changes to ecosystems occur faster, perhaps especially when humans are involved. In the last chapter, we saw that beavers are ecosystem engineers, altering their environment for their own needs. This then has dramatic effects on the whole ecosystem (more ponds, more and different biodiverse organisms). We humans are also

This image shows the interface between city, agriculture and forested areas. Human communities like this one must learn to live in balance with the surrounding ecosystems.

ecosystem engineers, adapting the environment to meet our needs. Just like any other living thing, we have physical needs, like shelter, food, and water. Sometimes meeting our own needs has unavoidable effects on the environment. For examples, the development of urban areas dramatically changes the environment, and diverting water from a river can affect nearby ecosystems.

Caring for other people and helping provide for their needs (homes, food, water) is an important part of our responsibility as Catholics, as a form of charity and service. But knowledge of ecology could help us do these important things *while also being good stewards of God's creation*. For example, in choosing a location to develop land for human use, we could measure biodiversity in each location, or estimate effects on populations of important or unique species, and choose the best location. A big part of environmental stewardship is being aware of the effects of our actions and being intentional about our choices.

There are also times when human activity may harm the environment in unintentional or accidental ways. Loss of habitat can remove available space for certain species, or pollution could disrupt an ecosystem. As one of the more common examples, oil spills from ships transporting oil or leaks in pipelines harm the environment. But after these spills or other disturbances, we can also try to restore ecosystems to their former function. This process is called **ecological restoration**. Restoration includes removing the source

A crude oil spill washes up on a beach. These sorts of accidents can have dire consequences on ecosystems and the organisms that call them home.

Helpful Microbes

One tool for ecological restoration is the discovery or development of microbes (through genetic modification) that can remove pollutants from the environment. Some amazing communities of bacteria can, collectively, metabolize (or "eat") oil within the environment. Other microbial species or communities may be able to break down plastics. We have also discovered that some species of insect larvae can eat Styrofoam! Cutting down on things like plastics and other sorts of trash can greatly affect our environment.

of disturbance, reintroducing key species, and monitoring the community over time.

Sometimes knowledge of ecosystems and ecology can also help us avoid unintentionally causing harm to humans and the environment. One of the most famous examples of ecological restoration is the Kissimmee River in Florida. In the 1940s, flooding from hurricanes was an issue, and so we made a plan to "canalize" the river (make it straighter, less meandering) and to add in structures along the way to hold flood water. But this project resulted in population reductions in many birds, fish, plants, and other species. It also increased pollution and other issues downstream (wetland plants are good at helping filter pollution, and these were also lost). So it ended up causing more harm than good, leading the United States Army Corps of Engineers to slowly restore the river to its former course. The riverside communities slowly recovered, and reestablished some ecosystem services they provide for us (fish and prawns to eat, drawing pollutants out of the water, etc.).

It may seem like you or I can do very little to help protect our ecosystems and all of God's creation (perhaps especially so as a young person). But all good things start small. Simply look at Christ's Church—what started as a small gathering of twelve men in the upper room of a humble home has spread throughout the entire world, now with millions upon millions of followers. This should remind us of Jesus' parable about the Kingdom of Heaven being like the mustard seed (Mk 4:30–32). Take time to discover what little daily actions you can take to protect our environment. If we all do this, we can make a world of difference.

FOUNDATIONS REVIEW

✓ All of the interactions we have studied thus far scale up and come together to provide an interconnected ecosystem. The "system" part of the word ecosystem reminds us that it operates as a *unit*, and changes to one aspect of the ecosystem affect other parts. By studying these changes we can better understand how ecosystem services help us (how the environment can give us things we need to survive).

✓ Ecological succession refers to how an ecosystem changes over time. These changes can be swift or take hundreds or thousands of years, and they can be complete changes to the ecosystem as a whole or smaller changes within the larger ecosystem.

✓ Human activity is one way ecosystems can change, and even change dramatically. As we seek to be stewards of God's creation and enjoy the fruits of the environment, we must be conscious of how our actions can affect nature.

Conclusion

To wrap up our book, I want to emphasize the connections between ecosystems, scaling up even beyond the level of the ecosystem. One last key term is **landscape**—a connection of various patches of different habitats and ecosystems. As you look out over the landscape from a mountaintop, it may look like a patchwork mosaic of forests, prairies, rivers, and wetlands (or perhaps other biomes, depending on where in the world you are). These ecosystems are not independent from each other, but rather are interconnected via nutrient cycling, the water cycle, the movement of energy, and by organisms that might move through multiple habitat types. Studying the ecology of God's creation is simply awe inspiring, with so much to discover at each level of study—from individuals to populations, or from ecosystems to landscapes (and eventually Earth's entire biosphere!). Hopefully this book illuminates some of the interconnectedness of nature and starts you on a path of discovering more about our wonderful world!

MIGRATING ELEPHANTS IN THE AFRICAN SAVANNAH

AMAZING FACTS ABOUT ECOSYSTEMS AND SPECIES

- *Eco* comes from the Greek word *oikos*, which means "household." In literal terms, then, ecology would be the study of the home or household—which makes sense if we consider our home to mean our planet—Earth.

- Ecologists study organisms and their relationships with the environment and with other living things. They may study an organisms behavior, physiology, lifespan, food, natural predators, diseases and more. Topics include how individuals respond to other living organisms or to their environment (temperature, moisture, light, chemicals, etc.). For example, we could ask how well a blue jay remembers where it buried acorns. Or how does increasing average temperature affect the lifespan of a pika?

- Some species have interesting names, including one jumping spider—*Indomarengo chavarapater*—named after St. Kuriakose Elias Chavara from India!

- When we study nature, we see that organisms are adapted to the environments and habitats they live in—that is, they have traits that match the challenges around them. For example, animals in the arctic have behavior and physiology that helps them in the cold, while desert animals have traits for hot environments.

- The Sahara Desert has daily high temperatures over 100 degrees F for several months during the summer. In contrast, the tundra in Canada is consistently below 0 degrees F during the winter—even reaching below -60 degrees F from time to time! We wouldn't expect animals, plants, or fungi to be able to survive in both these places—the swings in temperature are just too dramatic. Instead, organisms are adapted to deal with either hot or cold conditions, depending on where they live, and have traits that help them survive.

- Sweating is one way that many mammals cool off, including us. The body releases sweat, and as it evaporates off the skin, that transition of water from liquid to gas decreases the temperature of the skin.
- Some kangaroo species living in the hottest parts of Australia lick their forearms to add more cooling—the saliva evaporates and cools the area like sweat evaporating. Moreover, there are many blood vessels under the forearm, so the blood there cools off before returning to the heart and circulating elsewhere. In this way, licking their forearms can help them cool off their whole body!
- You probably know that fur for mammals and feathers for birds help them keep warm, but some animals have layers of blubber (a type of fat) that also provide insulation. Elephant seals (*Mirounga angustirostris*), which sometimes swim in cold-water environments, have bodies that are up to 40 percent fat!
- Some animals do not keep themselves warm, but use outside sources of heat to warm themselves up—these are called ectotherms (the prefix *ecto* comes from the Greek *ektos*, meaning "outside," and *therm* means "heat," so "heat from outside"). Birds and mammals that use their own body heat to stay warm are endotherms (*endo* from the Greek for "inner" or "inside"). But ectotherms that can't warm themselves from the inside might bask in the sun to warm up, which is why you see lizards and snakes lounging in the sun on warm rocks or bricks. Ectotherms use less energy because they don't use their own energy to raise their body temperature, but they may be limited in where they can live because they must depend on outside factors to stay warm.

- Wood frogs live throughout Canada and Alaska because they can survive even frozen solid in winter (like a frog Popsicle) and simply thaw out in warmer months!
- Various species of fungi use chemicals in their cells which act as antifreeze. Some plants, like pine trees, move water outside their cells—it sits between cells instead—and so when ice forms it doesn't do as much damage to the plant. This is one reason why coniferous trees (evergreens), such as pines, are found in colder environments.

- Some animals hibernate and some plants enter a state of dormancy during winter months to survive the cold.

- Some desert plants have thin leaves to prevent damage from the strong sun. Others have spines (a type of modified leaf) which are white or fuzzy (like some cactuses such as the teddy-bear cholla). These reflect sunlight away from the plant, preventing the energy in the sunlight from being absorbed by the plant. This effect is similar to how we can experience a difference in temperature based on our clothing color—wearing black clothing in summer absorbs more sunlight and could cause us to feel warmer.

- Like animals, plants also use evaporation as a cooling mechanism. Yes, in a way, plants sweat! They release water through small pores in the leaf tissue called stomata; the evaporation of this water cools the plant during periods of high temperatures.

- The cells and bodies of organisms—whether plants, animals, fungi, or other living things—are made up of approximately 60 to 90 percent water; the exact figure differs among the species.

- One biome is called the tropical dry forest, which has dramatic swings from its rainy to its dry season. Up to seventy-five inches of rain can fall during the rainy season and as little as two inches in the dry season!

- Some desert plants only open stomata at night, which provides the carbon dioxide needed for photosynthesis without losing as much water to evaporation (which would occur more in the heat of the day). Many desert plants also have water storage capabilities—cactuses and other succulent plants store water in fleshy leaves or stems. These traits allow plants to hold onto the water they already have.

- Some plants are well-adapted to dry conditions because they can gather dew or water from the air. The Namib Desert is an extremely dry region along Africa's southwest coast. Because of its proximity to the coast, though, fog rolls over the area often enough for plants to use it for their water source. As water condenses from the fog onto grasses or other plants, it is channeled to drip down toward the base of the plant, where the roots can absorb it.

- Animals can collect dew and fog, too. An estimated fifty species or more collect water from the fog in the Namib Desert. One beetle species (*Lepidochora kahani*, also called the flying saucer trench beetle) acts as a little architect, building dew collection traps out of trenches dug into the sand. In the United States, the Texas horned lizard (*Phrynosoma cornutum*) has a specialized set of tiny channels or grooves covering its scales. When the lizard positions its body in the right way (head down), gravity and these channels direct the dew or other moisture on the lizard toward its mouth so that it can take a drink. Remarkable!

- Various desert mammals have fewer sweat glands than their non-desert counterparts—resulting in less water lost via evaporation, though this also means less cooling via sweat (instead, they must find other ways to cool off, like being active at night).

Ecosystems Fun Fact: Every continent except Antarctica has grasslands.

- The aptly named fat-tailed dwarf lemur stores fat in its tail to use as an energy source when it hibernates. Rather than avoiding cold, though, the lemur hibernates during the dry season in its tropical dry forest habitat. Plants in this forest also rest during the dry period, producing flowers and fruits when water is more available in the rainy season.

- Only certain plants thrive in wetlands and areas that undergo persistent or periodic flooding. Bald cypress (*Taxodium distichum*) is a type of coniferous tree that produces knobby "knee" structures from their roots that emerge from the ground. Scientists think this may help their roots continue to receive oxygen during flooding, and also may help stabilize the plant in the soft, water-logged soil.

- Adaptation can, perhaps confusingly, refer to two different things. It can be used to describe a process of change over time, but also physical traits or characteristics an organism might have because of those processes which helps it survive in its environment. So an adaptation (a trait) can be the result of adaptation (the process).

- The process of adaptation helps organisms become "better fit" within their environments. This often happens through natural selection, where individuals with the best success pass on beneficial traits to their offspring.

- All life on earth has the instructions on how to build it "written" into the DNA contained in its cells. In this way, we say that DNA is an information-carrying molecule. Each of our own cells—skin cells, lung cells, liver cells, et cetera—has our entire set of DNA instructions within them.

- DNA and the environment can affect traits. For example, tall parents are more likely to have taller children, but things like access to good nutrition also help children grow tall. So children are not *exactly* as tall as their parents, but instead development, nutrition, and other factors play a role. One estimate suggests that height is 80 percent genetic—so overall, children are still *more likely* to be tall if mom and dad are both tall.

- Human beings often engage in artificial selection with agriculture and livestock, for example breeding cows that produce more milk or corn that has larger ears with more kernels.

- Disease rates increase as the population increases. This is because individuals interact with other individuals more often as they compete for food and space, and this provides opportunities for bacteria and viruses to move from individual to individual and more quickly spread through the population.

- Koalas are often used as an example of specialists because they only eat leaves from the eucalyptus tree. So koalas have the narrow diet of a specialist, unlike other species that eat a wide range of foods. Much different than this would be another marsupial mammal, the wombat, which eats various grass species and other plants on the ground. Wombats are generalists, eating a broad, general assortment of foods. For animals foraging in the wild, being a specialist often allows greater competitive ability, at least for those foods they specialize on (sort of like getting really good at one particular skill). But generalists can avoid competition by opportunistically taking whatever food is most available.

- A classic example of niche partitioning in ecology is the species of *Anolis* lizards in the Caribbean. Different species specialize in using different parts of their habitat to live and forage—along the ground, on shrubs, on tree trunks, or in treetops. On the islands this "space sharing" pattern is repeated, but often with different species filling an individual niche. For example, each island has a ground-dwelling species, but different species can occupy the ground-dwelling niche. If a new species is introduced to the island, it would have the most success (highest fitness) if it fills an open niche, and so eventually all niches are occupied. The pattern of foraging in different areas reduces competition and allows coexistence of multiple species of lizards. It's truly remarkable how adaptive animals can be!

- Camouflage is coloration to help an animal blend in with its surroundings, like background matching. Disruptive coloration is using splotches, spots, or stripes to break up the shape of the animal's body—predators are less likely to spot you if you don't look like an animal!

- In addition to camouflage or disruptive coloration, organisms could also use mimesis to look like something uninteresting or non-threatening, like a leaf or twig, or mimicry to look like something dangerous—like caterpillars that have eyespots to resemble a snake!

- Some organisms have chemical defenses—compounds in their tissues that are distasteful or even toxic. Various species of milkweed are one example of this, with bitter compounds that are poisonous (cardenolides, which interfere with the function of cells—cattle can even die from eating milkweed).

- Predators can also use camouflage to sneak up on or surprise their prey. For example, some mantises and spiders camouflage themselves as flowers to catch insect prey.
- Monarch butterflies lay eggs on milkweed, the plant mentioned a moment ago, and the caterpillars are able to sequester (sort of like "save up") the toxic chemicals inside their own bodies. This means they are specialized to eat the poisonous plant and, instead of being harmed, become poisonous themselves!

- Predators and prey affect each other's fitness, so many traits and adaptations are responses to the pressures from other species.
- Bats use what is called echolocation to hunt insects. This is a way of locating prey (food) by using sound waves since they have poor vision and often hunt at night. The bats emit a sound wave which will reverberate back an echo—if that echo is disrupted or changed by the presence of something in the path of the wave, the bat knows something is there, in this case an insect. But some species have evolved to make clicking sounds to confuse echolocating bats. The added noise makes it harder for bats to locate the moth.
- Some crickets in Hawaii lost the ability to sing, or chirp. Parasitoid flies lay eggs on male crickets, and their larvae eat them from the inside out. Parasitoids are able to find the males when they sing to attract females. On some islands of Hawaii, crickets have stopped singing in the last few decades. This helps them avoid being detected by parasitoids.
- One fascinating example of cooperative hunting seen in nature is the brown-necked raven (*Corvus rufficollis*), which has been observed working together in groups to catch lizards. First, two ravens swoop towards the burrow of the lizard, cutting off its escape. Only then do a few others fly down to catch the

Species Fun Fact:
Cactuses can live without a drop of rain for several months because of adaptations they have developed to store water for long periods of time.

lizard. The whole enterprise seems planned out, and each individual seems to know its job. The burrow-blocking ravens only join their flock mates to share in the meal once the hunt is completed.

- Fish have been shown to cooperate in fascinating ways, including by investigating dangerous looking shapes together. In experiments, scientists will place a large predator-like shape in an aquarium, something that casts a looming shadow. The small "test fish" in the school inch forward a small distance at a time to check out the possible danger and will keep moving up as long as others in the school match their behavior. This means that fish only work together to check out danger if others are willing to help out!

- In biology, the word trophic, or word endings of -troph, indicates something about food. The words come from the Greek word *trophikós*, which means "pertaining to food." Trophic levels show where an animal is in a food diagram. Other quick examples include autotrophs, which are organisms like plants that generate their own food ('auto' means 'self'; troph = food), while heterotrophs eat other organisms (hetero = another, so eats another to get food).

- Because energy transfers up the food chain, but does so inefficiently, there are always more producers than consumers, and so this bottom-up effect is always present. In fact, the food webs found in communities across the world tend to have similar ratios of each trophic level. For example, one study showed that prey species tend to outnumber predators at about a 2.5 to 1 ratio (2 to 3 prey species for each predator species present in a food web).

- An interesting effect of the inefficiency of energy being transferred up the food chain is that there are limits to how many links can be in a food chain. The average number of links (from producer to the top predator, counting each step in-between) is about five, and they rarely, if ever, get to ten.

Species Fun Fact:
Bees can fly up to 20 mph!

- A good example of an apparent competition comes to us from an experiment that studied the effects of a mustard plant that was accidentally introduced into native plant communities in the United States. It seemed that the mustard could outcompete native grass species, taking over the area. But researchers studying the system showed that the biggest effect wasn't competition between plants; instead, mustard provided a safer shelter for mice to hide from predators. In turn, this increased the population of mice, which prefer to eat grass seed. So it looked like the mustard plant was driving out the native grasses, but it was not doing this directly—only because the mice were surviving longer and thereby eating more grass seeds.

- Bees pollinating crops, bacteria removing waste from the environment, and wetland plants preventing flooding are all good examples of ecological services (ways humans benefit from nature).

- Some animals are ecosystem engineers, meaning they can have a large impact on their community by altering their own habitat, like beavers which build dams. This also has the effect of creating habitat for all sorts of plants and animals that utilize the ponds the beavers make—whether aquatic critters living in the pond, or those that use the pond for food or drink. Other examples of these "engineers" include prairie dogs which create tunnel systems that animals can use as shelter and also affect plant life; or woodpeckers which make nesting holes in trees, which are then later used by wood ducks and other species (who can't make their own nesting holes).

- A research project at Utah State University has some unique research partners—beavers! The goal is to help restore stream health in Western states. Some small streams have started drying up in summer months, but beaver dams built along the course of a stream can help keep water in it year-round (storing it up behind ponds). Reintroducing beavers, or having students mimic the building of beaver dam structures, can result in a reliable source of water for wildlife and livestock!

- One tool for ecological restoration is the discovery or development of microbes (through genetic modification) that can remove pollutants from the environment. Some amazing communities of bacteria can, collectively, metabolize (or "eat") oil within the environment. Other microbial species or communities may be able to break down plastics. We have also discovered that some species of insect larvae can eat Styrofoam! Cutting down on things like plastics and other sorts of trash can greatly affect our environment.

GROTTO FALLS, SMOKY MOUNTAINS, TENNESSEE

KEY TERMS

Abiotic – *Chapter 1*: Means non-living (the prefix "a" means non and bio = life, so abiotic = non-living); abiotic factors refer to non-living things which affect organisms in their environment, like temperature or rainfall.

Adaptation – *Chapters 2 & 4*: *Adaptation (process)* – Ways in which organisms become more suited to their environment over time. *An adaptation* (a physical or behavioral trait) – a trait that is the result of an adaptive process (see *Adaptive traits*).

Adaptive traits – *Chapter 4*: The traits that result through the process of adaptation.

Altruism – *Chapter 9:* When an organism helps or benefits another individual at a cost to itself, with no clear benefit or immediate reward to itself.

Alarm calls – *Chapter 9*: When animals use audible calls to signal danger and bring others to help fend off a predator attack; various species of birds and mammals use alarm calls to warn others (and other species may use something other than sound).

Anthropomorphism – *Chapter 9*: The practice of giving human-like characteristics to non-human animals (like saying an animal is "sad").

Apex predator – *Chapter 10*: A predator that sits at the top of the food chain and has no predators that target it.

Apparent competition – *Chapter 10*: When species that do not directly compete still affect each other indirectly through their relationship with another species, such as having a common predator.

Artificial selection – *Chapter 5*: The process by which human beings select which traits in an organism they wish to pass down from one generation to the next (as opposed to natural selection).

Biodiversity – *Chapter 11*: The amount of diversity (variation in species) found in a given biological community.

Biological fitness – *Chapter 5*: The measurement of "success" or "failure" of how well an organism survives and reproduces in its environment, relative to others in the population.

Biomass – *Chapter 11*: The added sum weight of living things in a given area. Scientists may measure the total biomass of an area, or of individual species (for example, which tree species has the most biomass in this forest?).

Biomes – *Chapter 1*: Types of ecosystems found in particular geographic regions, defined primarily by temperature and rainfall patterns.

Biosphere – *Chapter 1*: The entirety of all the ecosystems on Earth; our planetary ecosystem.

Biotic – *Chapter 3*: Living things or organisms (as opposed to abiotic, non-living things).

Blubber – *Chapter 2*: A type of fat that some animals have to insulate them from the cold.

Camouflage – *Chapter 8*: Also called cryptic coloration, it is a defensive or hunting mechanism that allows organisms to blend in with their environment.

Carrying capacity – *Chapter 6*: The number of individuals of a given species that a habitat can support (that is, "carry") based on space and resources (like food and water).

Chemical defenses – *Chapter 8*: Traits an organism has to protect itself using chemical compounds, to make it less desirable to eat or even dangerous.

Remember:
The more diverse an ecosystem is, the better it can thrive and overcome both biotic and abiotic challenges.

Coevolution – *Chapter 8*: When two species evolve in response to one another, for example, predator and prey develop counter-adaptations in response to one another.

Commensalism – *Chapters 7 & 9*: A type of symbiosis where one species benefits and the other is more or less unaffected.

Community – *Chapter 1*: The collection of populations of various species that all interact with one another in a particular area.

Competition – *Chapter 7*: The way organisms fight or compete with one another for food and other resources in their environments.

Competitive exclusion – *Chapter 7*: When one competitor excludes the other from the local area by dominating them directly (like preventing access) or securing the resources they both target.

Complexity (of an environment) – *Chapter 11*: An environment that has a high variation in species and niches (for example, patches of sun and shade, different plant types, soil types, etc.).

Consumers – *Chapter 10*: Animals that eat the producers (plants) at the base of the food chain.

Cooperation – *Chapter 9*: When members of the same species or different species engage in behavior in which both benefit.

Counter-adaptation – *Chapter 8*: When one organism finds a way to counter another organism's advantage; for example, as they engage in a predator-prey relationship.

Density-dependent factor – *Chapter 6*: A factor in a given environment that affects populations based on how densely populated the environment is with certain organisms (effects *depend* on the density).

Density-independent factor – *Chapter 6*: A factor in a given environment that affects populations that is *not* based on how densely populated the environment is with certain organisms (affecting the population *independently* of how dense it is).

Dew – *Chapter 3*: Condensation of the water in the air onto the ground and other surfaces.

Disruptive coloration – *Chapter 8*: A form of camouflage that uses stripes or spots to break up the outline of an animal.

Diversity index – *Chapter 11*: A measure of how rare or abundant various species are in a given biological community.

Dominant species – *Chapter 11*: The most abundant and influential species found in a given biological community.

Dormancy – *Chapter 2*: A hibernation-like state of rest some plants and fungi can enter during winter.

Ecological (biological) community – *Chapter 10*: The set of biological organisms that occupy and interact in a particular area or environment.

Ecological niche – *Chapter 7*: The role an organism plays in the ecosystem, in its environment and its habitat requirements (space, temperature, nesting sites, etc.), as well as the interactions it has with other species.

Ecological restoration – *Chapter 12*: Efforts human beings can take to restore ecosystems after they have been disturbed or their balance upset in some way; includes such actions as removing the source of disturbance, reintroducing key species, and monitoring the community over time.

Ecological succession – *Chapter 12*: The changes that occur in an ecosystem over time.

Ecologists – *Chapter 1*: Scientists who study ecology.

Ecology – *Chapter 1*: The study of living things and their relationship—or "interconnectedness"—with each other and their surrounding environment.

Ecosystem – *Chapter 1*: Refers to the biological communities living in an area and how these communities (with all their populations of organisms) are affected by abiotic factors and the physical environment.

Ecosystem engineers – *Chapter 11*: Organisms that can alter their own habitat (for example, beavers building a dam) and even create habitats for other organisms.

Ecosystem function – *Chapter 12*: The set of processes that are necessary to maintain the living systems in an ecosystem.

Ecosystem services – *Chapters 11 & 12*: All the ways (goods and services) in which humans benefit from nature.

Ectotherms – *Chapter 2*: Animals that cannot keep themselves warm but rather must use outside sources of heat to warm themselves up (the prefix *ecto* comes from the Greek *ektos*, meaning "outside," and *therm* means "heat," so "heat from outside").

Emigration – *Chapter 6*: When individuals leave a population in one geographic region and join another in a different region.

Endangered species – *Chapter 6*: An organism that is in danger of going extinct.

Endotherms – *Chapter 2*: Animals that can use their own body heat to stay warm (*endo* from the Greek for "inner" or "inside").

Environmental heterogeneity – *Chapter 11*: The presence of high variation in species and niches (complexity) in nature.

Evolution – *Chapter 4*: A change in the characteristics (or traits) of a population of organisms over time or over generations.

Exploitation – *Chapter 8*: When one organism exploits (uses) another for a source of food or energy.

Facultative mutualism – *Chapter 9*: A kind of mutualism where the two organisms engage in an opportunistic relationship but could survive without the other and easily switch "partners."

Food chain/web – *Chapter 10*: The interactions and connections in nature of various species eating other species.

Generalist – *Chapter 7*: An organism that does not specialize in any one set of resources; for example, animals with a broad diet.

Gene – *Chapter 5*: A segment of DNA (deoxyribonucleic acid) which contains the instructions for how to build a particular protein, or controls the development of a particular part of an organism. Each organism has many genes, each containing "coding" for different traits.

Ecosystems Fun Fact: Some tropical dry forests can get up to seventy-five inches of rain during the rainy season and as little as two inches in the dry season!

Genetic drift – *Chapter 4*: Describes random fluctuations in the traits of a population over time, and it is one type of change that does not necessarily lead to better adapted populations.

Guild – *Chapter 10*: Similar to a trophic level; a set of organisms that exploit the same or similar resources.

Herbivore – *Chapter 8*: An animal that eats plants.

Heritability/Heritable – *Chapter 5*: The passing down of traits from one generation to the next. A heritable trait is one that is passed down from parent to offspring.

Hibernation – *Chapter 2*: A mostly dormant, sleep-like state some animals enter after storing up fat to save energy during colder months; a deep state of physical resting.

Host – *Chapter 8*: The organism that harbors the parasite—essentially it is the home for the parasite.

Humidity – *Chapter 3*: Water vapor in the atmosphere.

Immigration – *Chapter 6*: Individuals joining a population from another geographic region.

Inquiline – *Chapter 7*: A species that lives in the home of another species without harming it.

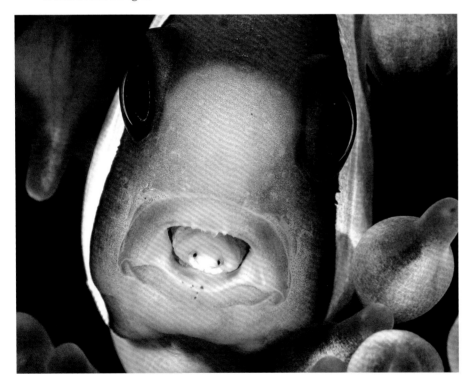

Interference competition – *Chapter 7*: When individuals compete directly for a specific resource.

Interspecific competition – *Chapter 7*: Competition between individuals of two or more different species.

Intraspecific competition – *Chapter 7*: Competition between two or more individuals within the same species.

Invasive species – *Chapters 6 & 11*: Any animal or plant not native to a given area that has a destructive effect (causing ecological, economic, public health or other harm).

Larvae – *Chapter 8*: The immature or juvenile form of an insect.

Landscape – *Conclusion*: a connection of various patches of different habitats and ecosystems.

Local adaptation – *Chapter 4*: Adaptations within the same species that vary from population to population depending on what environment they live in.

K-selected – *Chapter 6*: A classification of species known for having few offspring but that provide intensive care for their young.

Keystone species – *Chapter 11*: A species that has an especially large or important effect on their community; the effect it has on the community is greater than what is expected given how abundant it is (rare species can often be keystone species).

Mimesis – *Chapter 8*: When an organism can look like something non-threatening in the environment, like a twig or leaf; a form of cryptic coloration.

Mimicry – *Chapter 8*: When an organism can look like something threatening or dangerous—like caterpillars that have eyespots to resemble a snake.

Mutualism – *Chapter 9*: A type of cooperation seen in nature when two organisms engage in a mutually beneficial relationship and do something helpful for the other.

Myrmecophile – *Chapter 7*: A type of inquiline that lives with ants (*myrmecophile* translates as "lover of ants").

Natural selection – *Chapter 5*: The process by which populations of living organisms adapt and change; those individuals with desirable traits have higher fitness (higher survival, more offspring), and so these desirable traits will increase in the population (more offspring born with those traits), leading the population to become better suited to its environment.

Niche partitioning – *Chapter 7*: When organisms avoid competition by decreasing the overlap of their ecological niche; for example, two seed-eating bird species could each specialize on a particular type of seed, decreasing competition between them.

Obligate mutualism – *Chapter 9*: A kind of mutualism where the organisms are entirely dependent upon one another for survival.

Parasite – *Chapter 8*: An organism that lives on or inside a host organism, feeding on the body of the host, typically without killing it.

Parasitoid – *Chapter 8*: A combination of a predator and a parasite—an organism (often an insect) whose larvae kill and eat a host organism.

Pathogen – *Chapter 8*: A disease-causing organism (such as viruses and bacteria).

Physiology – *Chapter 2*: All of the parts of an organism and the normal functioning of those parts (for example, muscles, blood vessels, and nervous systems in animals, or leaves, stems, and roots in plants).

Population – *Chapter 1*: A group of individuals of the same species within a given area.

Population biology – *Chapter 6*: The study of populations and how they change over time (for example, increasing or decreasing in abundance).

Population dynamics – *Chapter 1*: Descriptions of how and why population size changes over time.

Predation – *Chapter 6*: One animal preying on another.

Predator – *Chapter 8*: An organism that kills and eats other organisms—typically animals that eat other animals.

Species Fun Fact: Eco comes from the Greek word oikos, *which means "household." In literal terms, then, ecology would be the study of the home or household—which makes sense if we consider our home to mean our planet—Earth.*

Prey – *Chapter 8*: The organism a predator kills and eats.

Producers – *Chapter 10*: Plants and other organisms that occupy the base of the food chain by conducting photosynthesis (turning sunlight into food energy), introducing energy into the food chain.

Productivity – *Chapters 11 & 12*: The amount of plant growth—the generation of biomass—in a given biological community; essentially, how efficiently producers turn light energy into growth.

R-selected – *Chapter 6*: A classification of species that is known for having many offspring but that provides little to no parental care; offspring tend to be smaller in size but reach maturity quickly.

Resource competition – *Chapter 7*: Also known as *scramble competition*—when two or more individuals compete indirectly for resources.

Scramble competition – *Chapter 7*: See *Resource competition*.

Specialist – *Chapter 7*: An organism that specializes on a specific set of resources; for example, having a very narrow diet of one or a few food types.

Species – *Chapter 1*: A distinct type of organism; or a group of similar living things that is distinct from other such groups.

Species evenness – *Chapter 11*: A measure of biodiversity that describes how equally balanced the various species are in number in a given biological community.

Species richness – *Chapter 11*: A measure of biodiversity that provides a count of how many species are present in a given biological community.

Structural defenses – *Chapter 8*: Physical traits an organism might have to protect itself.

Remember:
A mutualism is a type of cooperation seen in nature when two organisms engage in a mutually beneficial relationship and do something helpful for the other.

Symbiosis – *Chapter 9*: When two organisms interact with each other and live together in close association.

Trophic cascade – *Chapter 10*: When a single change in a community ripples through the food web and causes larger changes; can come from the top-down (from predators down) or from the bottom-up (plants/producers upwards).

Trophic level – *Chapter 10*: A description or categorization of where certain species sit on the food chain.

Water balance – *Chapter 3*: The amount of water an organism is composed of and the daily effort to maintain that level.

IMAGE CREDITS

Front cover 3d isometric Ecosystem, SERGEYTEDBEAR / Cardinal, Bonnie Taylor Barry / rhinoceros, male sika deer, Anan Kaewkhammul / American alligator, Coahuilan Box Turtle, fivespots / Earthworm, galitsin / white clouds, warat42 / Jewel beetle, Photoongraphy / Greylag Goose, Resul Muslu / Reaching hands from The Creation of Adam of Michelangelo, Freeda Michaux © Shutterstock.com

p1 The head and neck of a female Roe deer just visible above the rough vegetation © SciPhi.tv, Shutterstock.com

pII Spring sunrise over a river bank with morning fog © Sergey Nesterchuk, Shutterstock.com

pV Space background with many stars © Sundays Photography, Shutterstock.com

pVI Green sea turtle (*Chelonia mydas*) © Rich Carey, Shutterstock.com

pVIII Wood Frog (*Rana sylvatica*) © Michiel de Wit / Carolina K. Smith MD, Shutterstock.com

pIX Ecosystem of trees and rocks © Artiste2d3d, Shutterstock.com

pX European Bee-eater (*Merops apiaster*) © sokal, Shutterstock.com

pXII-p1 Yosemite Valley, California © Jon Chica, Shutterstock.com

p2 Large Hawk in flight © Stephen Mcsweeny, Shutterstock.com

p2 Gray common house mouse © dwi putra stock, Shutterstock.com

p3 White tailed buck deer in midwest farm country © Tom Reichner, Shutterstock.com

p3 Mule Deer buck, Montana © Tom Reichner, Shutterstock.com

p4 Animals and plants © A7880S, Shutterstock.com

p4 Biosphere diagram © Platon Anton, Shutterstock.com

p5 Wadi Mujib Biosphere Reserve, red rock ravine gorge with river. Jordan water stream with blue sky. Wadi Mujib lowest nature reserve landscape, with a spectacular array of scenery near Dead Sea. © Ondrej Prosicky, Shutterstock.com

p5 Tropical jungle © Teo Tarras, Shutterstock.com

p5 Winter landscape in Austrian Alps © Olha Sydorenko, Shutterstock.com

p5 Lumphini Park, Bangkok, Thailand© PICONEST, Shutterstock.com

p6 Green moss and rocks © xpixel, Shutterstock.com

p7 Saint Francis preaching to the birds, 1913 / Reid, Stephen (1873-1948) / English / © The Stapleton Collection / Bridgeman Images

p8-9 Lion resting in the shade of a tree © BZ Travel, Shutterstock.com

p10 Sahara desert © takayuki, Shutterstock.com

p10 Alaskan tundra with Philip Smith mountains in the Arctic National Wildlife Refuge © Troutnut, Shutterstock.com

p10-11 Gaggle of Canada geese in formation flight © DavidEdwards8, Shutterstock.com

p11 Jaguar in tropical rainforest at night © Anan Kaewkhammul, Shutterstock.com

p12 Close up of Australian red kangaroo licking its paw © AdamOliver, Shutterstock.com

p12 Snow monkeys hugging and sheltering each other from the cold snow © Ro_Ma_Li, Shutterstock.com

p12 Southern Elephant Seal pups (*Mirounga leonina*) on a sandy beach on Sealion Island in the Falkland Islands. © Jeremy Richards, Shutterstock.com

p13 Arctic fox in Canadian arctic © outdoorsman, Shutterstock.com

p14 Flock of Canada geese flying © rck_953, Shutterstock.com

p15 Saint Paul Writing. Date: 1520s. Source: National Gallery, London. Artist: Pier Francesco Sacchi (1485–1528) [public domain], via Wikimedia Commons

p15 Heimsuchung, Szene: Maria und HI. Elizabeth, HI. Nikolaus und HI. Antonius, Detail: HI. Anthony. Date: 1480-1490. Source: The Yorck Project (2002) 10.000 Meisterwerke der Malerei (DVD-ROM), distributed by DIRECTMEDIA Publishing GmbH. ISBN: 3936122202. Artist: Piero di Cosimo (1462–1522) [public domain], via Wikimedia Commons

p15 Desert landscape © Perfect Lazybones, Shutterstock.com

p16-17 Talamati waterhole, mixed game gather at a waterhole during a long drought. Kruger National Park, South Africa © WOLF AVNI, Shutterstock.com

p18 Silhouettes: Boy and girl, Chipmunk131 / Lizard, angel digital / fish, Hennadii H / jellyfish, EVKA / Dreaceana palm flower in flowerpot, Evelinna / bird, basel101658 / dog, pig, rabbit: Natalia Toropova © Shutterstock.com

p19 Miombo Forest, Kasungu National Park, Malawi. Source/Author: Dr. Thomas Wagner. License: (CC BY-SA 3.0), https://creativecommons.org/licenses/by-sa/3.0/deed.en

p19 Sunset Namib Desert, Namibia, South Africa. © kavram, Shutterstock.com

p20 Sagauro cactus flowers in spring time in the Arizona Desert © Ray Redstone, Shutterstock.com

p20 Early morning mist in a valley on the edge of the Namib Desert, the only moisture this area gets, Namibia © Stephen Barnes, Shutterstock.com

p21 A Lepidochora beetle, known as a trench digging beetle, in the Namib Desert, Namibia © Chantelle Bosch, Shutterstock.com

p21 The spiky Texas Horned Lizard (*Phyrnosoma cornutum*) © Matt Jeppson, Shutterstock.com

p21 Fat-tailed Dwarf Lemur (*Cheirogaleus medius*) 11 years old © Eric Isselee, Shutterstock.com

p22 Bald Cypress at the edge of a lake © Steve Bower, Shutterstock.com

p22 Brazilian Pantanal (jaguar) in a tree © Pedro Helder Pinheiro, Shutterstock.com

p23 Bald Cypress tress © Herschel Hoffmeyer, Shutterstock.com
p24-25 The Garden of Eden; in the background The Temptation (oil on oak panel) / Brueghel, Jan the Elder (1568-1625) / Flemish / Victoria & Albert Museum, London, UK / © Bridgeman Images
p26-27 Many butterfly species gather on a sunny day to find nutrients in the woods and on the water © furoking300, Shutterstock.com
p28 Peas plant with flowers and pods © Scisetti Alfio, Shutterstock.com
p29 Red ants © Penpitcha Phaungtham, Shutterstock.com
p29 Two Greater Egyptian jerboa (*Jaculus orientalis*) © jindrich_pavelka, Shutterstock.com
p29 Beaver walking on the ice © Lukas Jonaitis, Shutterstock.com
p30 Rare wild flower, Edelweiss flower (*Leontopodium alpinum*), growing high up in the mountains. © Kluciar Ivan, Shutterstock.com
p31 Swarm of mosquitoes near the reeds on a pond © VladKK, Shutterstock.com
p32 Landscape of a flood-meadow in Ukraine © Vadven, Shutterstock.com
p33 Group of snails © Oleksandr Lytvynenko, Shutterstock.com
p34 Portrait of the geneticist Johann Gregor Mendel, 1926 (pastel on paper) / Ezuchevsky, Mikhail Dmitrievich (1880-1928) / Russian / State Darwin Museum, Moscow / © Bridgeman Images
p36-37 Great Gray Owl resting in the crook of a tree. Winter in Manitoba, Canada © critterbiz, Shutterstock.com
p38 Oak forest © Jag_cz, Shutterstock.com
p39 3D illustration of DNA structure © Yurchanka Siarhei, Shutterstock.com
p39 Adult and puppy Border Collie © Erik Lam, Shutterstock.com
p40 Funny group of curious pigeons © Gallinago_media, Shutterstock.com
p42 Artificial selection infographic diagram; example brassica oleracea. © udaix, Shutterstock.com
p43 Holstein cow, 5 years old © Eric Isselee, Shutterstock.com
p44-45 Plateau Side Blotched Lizard in Antelope Island State Park, Utah © Cheri Alguire, Shutterstock.com
p45 Side-blotched Lizard, Western Subspecies (*Uta stansburiana elegans*) © Creeping Things, Shutterstock.com
p45 Common Side-blotched Lizard. Source: NPGallery [public domain], via Wikimedia Commons
p46-47 A crocodile hunting wildebeest during the great migration © Harry Collins Photography, Shutterstock.com
p48 Song Thrush (*Turdus philomelos*) nest with egg and hatchlings © Vishnevskiy Vasily, Shutterstock.com
p49 Adult black bear foraging in the forest in Cade's Cove, Great Smoky Mountains © Rick Grainger, Shutterstock.com
p50 A female orangutan in the rainforest of Borneo. © Sergey Uryadnikov, Shutterstock.com
p50 Green Sea Turtle swimming in Caribbean © Isabelle Kuehn, Shutterstock.com
p50 large bull moose walking through an open meadow in Grand Teton National Park © Tony Campbell, Shutterstock.com
p50 Zebra mingle with thousands of wildebeest on the banks of the Mara River during the great migration. In the Masai Mara, Kenya. Every year 1.5 million wildebeest make the trek from Tanzania to Kenya © Jane Rix, Shutterstock.com
p51 Wolves and brown bear fighting over a carcass © Peter Ivanyi, Shutterstock.com
p52 Diagram shows an exponential population growth that results in the J-shaped graph. The rapid growth causes an overshoot where the population is above the carrying capacity, this leads to a die off and unstable equilibrium as the population growth rate fluctuates above and below the carrying capacity. This overshoot and unstable equilibrium can lead to a degraded carrying capacity as the available resources have been destroyed by the overgrown population. Source/Author: Nchisick. License: (CC BY-SA 4.0), https://creativecommons.org/licenses/by-sa/4.0/deed.en
p53 Reindeers hamper traffic on the northern road in Lapland © Popova Valeriya, Shutterstock.com
p53 Burning tree © Alexyz3d, Shutterstock.com
p54 African Bush Elephant (*Loxodonta africana*) in Zimbabwe © Martin Pelanek, Shutterstock.com
p54-55 Butterfly eggs © Thammanoon Khamchalee, Shutterstock.com
p56-57 Strong stag beetle (*lucanus cervus*) lifting its rival over head with long mandibles during fierce fight © WildMedia, Shutterstock.com
p58 Close up of Red deer fighting during rutting season in UK © Giedriius, Shutterstock.com
p59 Carolina Wren © Christopher Unsworth, Shutterstock.com
p59 Eastern Towhee © Jayne Gulbrand, Shutterstock.com
p60 Bird feeder with bird; great tit (*Parus major*) © Nyvlt-art, Shutterstock.com
p60 Flying bird; titmouse (*Paridae*) © Bachkova Natalia, Shutterstock.com
p60 Flying bird; great tit (*Parus major*) © Potapov Alexander, Shutterstock.com
p60 Young squirrel © IrinaK, Shutterstock.com
p60 Hamster © VeryOlive, Shutterstock.com
p60 Chipmunk © Dimj, Shutterstock.com
p61 Spotted hyenas, (*Crocuta crocuta*), attacking a pride of lions, (*Panthera leo*) © MintImages, Shutterstock.com
p61 Leopard (*Panthera pardus*) © Eric Isselee, Shutterstock.com

p62 Camouflaged trail camera hidden on a tree under green moss © WildMedia, Shutterstock.com
p62 koala © Yatra4289, Shutterstock.com
p63 Wombat eating grass © Lenico, Shutterstock.com
p64 Two red kangaroos fighting © volkova natalia, Shutterstock.com
p65 Father Eric Wasmann, circa 1885. Source: Global Ant Project — World Ant Taxonomists [public domain], via Wikimedia Commons
p65 Weaver Ants (*Oecophylla smaragdina*) © frank60, Shutterstock.com
p66-67 Fly caught in a spiders web © Chrispo, Shutterstock.com
p68 A giraffe eating leaves off of an Acacia tree. © Haddonchris, Shutterstock.com
p68 A Tomato Hornworm with wasp eggs. A wasp has injected her eggs into this hornworm. When the eggs hatch into larvae, the caterpillar will be eaten. © Stephen Bonk, Shutterstock.com
p69 Mossy leaf-tailed gecko (Uroplatus sikorae) Montagne d'Ambre, Madagascar. two of a series of two showing the camoflage disguise using the dermal flap. Source/Author: Charles J. Sharp. License: (CC BY-SA 4.0), https://creativecommons.org/licenses/by-sa/4.0/deed.en
p69 Young skunk (*Mephitis mephitis*) © Nynke van Holten, Shutterstock.com
p70 Blowfish or puffer fish © FtLaud, Shutterstock.com
p70 Cactus © 2016_Darino, Shutterstock.com
p70 Indian porcupine (*Hystrix indica*) © PhotocechCZ, Shutterstock.com
p70 Hedgehog © David Muzik, Shutterstock.com
p71 Rough-skin Newt (*Taricha granulosa*) on a mossy log in the understory of a second growth forest of the Pacific Northwest. © Peter K. Ziminski, Shutterstock.com
p71 Eastern Gartersnake © Michiel de Wit, Shutterstock.com
p72 Echolocation for Bats. Source/Author: Shung [public domain], via Wikimedia Commons
p72 Common house mosquito (*culex pipiens*) © Danut Vieru, Shutterstock.com
p72 Grey long-eared bat (*Plecotus austriacus*) is a fairly large European bat. It has distinctive ears, long and with a distinctive fold. © Rudmer Zwerver, Shutterstock.com
p73 Eurasian Lynx (*Lynx lynx*) © Film Studio Aves, Shutterstock.com
p73 Snowshoe Hare changing from winter white to summer brown, near the Norris Geyser, northwestern Yellowstone National Park, Wyoming © Tom Reichner, Shutterstock.com
p73 Green cactus with red thorns © Tamara Kulikova, Shutterstock.com
p74-75 Saint Augustine in his office / Carpaccio, Vittore (c.1460/5-1523/6) / Italian / Scuola di San Giorgio degli Schiavoni, Venice, Italy / Luisa Ricciarini / © Bridgeman Images
p76-77 Red-billed oxpecker bird on impala's back, South Africa. © Danita Delimont, Shutterstock.com
p78 Wolf pack on the hunt in the forest © A. Kiro, Shutterstock.com
p78 Wild dogs on the hunt doe prey in moremi Game Reserve in the Okavango Delta in Botswana, Afrika © Benjamin B, Shutterstock.com
p79 Chaotic flight of snow geese in the late afternoon sun during spring migration at Middle Creek Wildlife Management Area. They are a species of goose native to North America. © Michael G McKinne, Shutterstock.com
p80 Several remora hitch a ride on a large lemon shark. © Greg Amptman, Shutterstock.com
p80 Lichens in various shapes, forms and textures on the surface of rocks. © Angelito de Jesus, Shutterstock.com
p80 Colorful soft corals at coral reefs under tropical ocean © Bangtalay, Shutterstock.com
p81 A fig wasp (*Philotrypesis sp.*, Family Pteromalidae) ovipositing into a fig (*Ficus burkei*) receptacle. Pietermaritzburg, South Africa. Source/Author: Alandmanson. License: (CC BY 4.0), https://creativecommons.org/licenses/by/4.0/deed.en
p82 Two-horned white Rhinoceros with oxpecker (tick bird) © Dmitry Rukhlenko, Shutterstock.com
p83 The Rescue of the Injured Man by the Merciful Samaritan (Evening) c.1856 (oil on canvas) / Schirmer, Johann Wilhelm (1807-63) / German / Neue Galerie, Kassel, Germany / © Museumslandschaft Hessen Kassel / Ute Brunzel / Bridgeman Images
p84-85 Spotted hyaena and white backed vultures in Kruger National park, South Africa. © PACO COMO, Shutterstock.com
p86 Food Chain diagram © BlueRingMedia, Shutterstock.com
p87 Brown bear © StockerrShmokerr, Shutterstock.com
p87 Coyote (*canis latrans*) © Jim Cumming, Shutterstock.com
p87 Side view of Red Fox, 1 year old © Eric Isselee, Shutterstock.com
p87 Brown rabbit on hay © Oleksandr Lytvynenko, Shutterstock.com
p88 Herd of Zebras grazing in the african bush. In the background a giraffe is eating leaves from an acacia tree © Willem Cronje, Shutterstock.com
p89 Aldo Leopold with quiver and bow seated on rimrock above the Rio Gavilan in northern Mexico while on a bow hunting trip in 1938. Source: Aldo Leopold trip to the Rio Gavilan: https://www.usda.gov/media/blog/2013/01/23/celebrating-life-and-legacy-aldo-leopold. Author: Pacific Southwest Region 5. License: (CC BY 2.0), https://creativecommons.org/licenses/by/2.0/deed.en
p90 Wolf and grey rabbit © photomaster, Shutterstock.com
p91 Princess Therese of Bavaria (1850-1925). Date: 1911. Source: https://badw.de/geschichte/chronik.html. Author: Friedrich August von Kaulbach (1850–1920) [public domain], via Wikimedia Commons
p91 Butterflies © aerogondo2, Shutterstock.com

IMAGE CREDITS

p92-93 Spotted fire salamander sitting on cep mushroom in autumn forest © Brum, Shutterstock.com

p95 Landscape view of the mangroves in Everglades National Park in Florida © WWyloeck, Shutterstock.com

p96 Wild Forests of the Carpathians are covered with thick fog after rain © Roman Mikhailiuk, Shutterstock.com

p97 17 year old Puma © Eric Isselee, Shutterstock.com

p97 A bull elk in autumn during the rut © Harry Collins Photography, Shutterstock.com

p97 Grand Teton National Park, Wyoming. Beaver (*Castor canadensis*) gnawing through an Aspen on pond shore. © Danita Delimont, Shutterstock.com

p97 praire dog in hole © Rob Francis, Shutterstock.com

p98 Beaver dam © SGeneralov, Shutterstock.com

p98 Large group of African fauna, safari wildlife © Eric Isselee, Shutterstock.com

p99 A black browed Albatross © robert mcgillivray, Shutterstock.com

p100-101 Aerial drone image of residential suburban houses backing onto a lush healthy green forest ravine © Alexander Gold, Shutterstock.com

p102 Cultivated plants in Germany, dried up in rows on dry, crusty soil. © Tanja Esser, Shutterstock.com

p102 Flooded cornfield in the midwestern United States © Suzanne Tucker, Shutterstock.com

p103 Jaws of a crane moving a log © djumandji, Shutterstock.com

p103 Root crops, carrots, parsley root, turnip, onion, garlic, Jerusalem artichoke, horseradish. © ulrich22, Shutterstock.com

p104 Ecological succession after wildfire in boreal pine forest (next to Hara Bog, Lahemaa National Park, Estonia): two photos of the same place, picture on the left taken one year and picture on the right two years after the fire. Source/Author: Hannu [public domain], via Wikimedia Commons

p105 Nature vs. architecture—an aerial view of Braila city Romania Eastern Europe © Craitza, Shutterstock.com

p105 A crude oil spill on the sand of a beach © A_Lesik, Shutterstock.com

p106 3d illustration highway road, lake, mountain and trees © Nomad90, Shutterstock.com

p107 Migration of elephants. Evening in the African savannah. © FOTOGRIN, Shutterstock.com

p108 Sand dunes in the Sahara Desert, Morocco © astudio, Shutterstock.com

p109 A water-snake sunbathing during early spring © DeepakkmrYadav, Shutterstock.com

p110 The teddy bear cholla is named for its furry "cuddly" appearance but is actually a densely spined plant © trotalo, Shutterstock.com

p111 A grey fox trots past a large sandstone boulder in the American southwest desert as it hunts for small prey animals at night © MelaniWright, Shutterstock.com

p112 White Swan Cygnets with Mother in the water © Denis Kuvaev, Shutterstock.com

p113 Gecko phyllodactylidae reptile with sand skin camouflage © Alexander Turovsky, Shutterstock.com

p114 Monarch butterfly laying her egg on the underside of common milkweed plant © K Quinn Ferris, Shutterstock.com

p115 Fish school on underwater coral reef in ocean © Rich Carey, Shutterstock.com

p116 A Bee hovering while collecting pollen from Pussy Willow blossom © Dave Massey, Shutterstock.com

p117 Grotto falls Smoky Mountains waterfalls © Dave Allen Photography, Shutterstock.com

p118 Killer Whale (*Orcinus Orca*) © Tory Kallman, Shutterstock.com

p119 Close-up on the fruits (*follicles aka pods*) of the common milkweed © krolya25, Shutterstock.com

p120 Large marsh grasshopper (*Stethophyma grossum*) © Silvia Dubois, Shutterstock.com

p121 Young lamb's ear leaves emerging from dormant plants in spring © BlueSnap, Shutterstock.com

p122 Panda bear eating bamboo © Hung Chung Chih, Shutterstock.com

p123 Small isopod entered the clownfish's mouth through the gills and ate the host's tongue. Ultimately, the isopod completely replaces the tongue of the clownfish and will live there. Although this is terribly gross, the anemonefish's life is not directly threatened by this. Source: A dentist nightmare. Author: Christian Gloor from Wakatobi Dive Resort, Indonesia. License: (CC BY 2.0), https://creativecommons.org/licenses/by/2.0/deed.en

p124 Starfish and sponge of the Mediterranean Sea © Adam Ke, Shutterstock.com

p125 Panther Chameleon ready to strike a praying mantis © lisdiyanto suhardjo, Shutterstock.com

p126 Field of dandelion flowers © Bilawal baledai, Shutterstock.com

p127 White rhino © Jason Prince, Shutterstock.com

p132 The Butterfly Pipevine Swallowtail (*Battus philenor*) emerging from it's chrysalis © IrinaK, Shutterstock.com

p132 Rocks surrounded by flowers © Artiste2d3d, Shutterstock.com

Back cover Common toad and her baby, European honey bee on a flowering plant, Eric Isselee / male sika deer, Anan Kaewkhammul / white clouds, warat42 © Shutterstock.com

The Foundations of Science introduces children to the wonders of the natural world in light of God's providential care over creation.

Too often we hear that science is in conflict with faith, but Pope St. John Paul II wrote that faith and science "each can draw the other into a wider world, a world in which both can flourish." *Foundations* seeks to spawn this flourishing in the hearts and minds of young readers, guiding them into a world that will delight their imaginations and inspire awe in the awesome power of God.

This eight-part series covers an extensive scope of scientific studies, from animals and plants, to the galaxies of outer space and the depths of the ocean, to cells and organisms, to the curiosities of chemistry and the marvels of our planet. Still more, it reveals the intricate order found beneath the surface of creation and chronicles many of the Church's contributions to science throughout history.

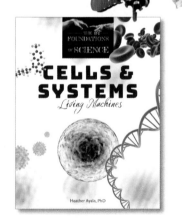

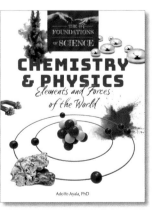